THEORY OF EXCITONS

SOLID STATE PHYSICS

Advances in
Research and Applications

Editors

FREDERICK SEITZ
*Department of Physics
University of Illinois
Urbana, Illinois*

DAVID TURNBULL
*Division of Engineering
and Applied Physics
Harvard University
Cambridge, Massachusetts*

The following monographs are published within the framework of the series:

1. T. P. Das and E. L. Hahn
Nuclear Quadrupole Resonance Spectroscopy, 1958

2. William Low
Paramagnetic Resonance in Solids, 1960

3. A. A. Maradudin, E. W. Montroll, G. H. Weiss
Theory of Lattice Dynamics in the Harmonic Approximation, 1963

4. Albert C. Beer
Galvanomagnetic Effects in Semiconductors, 1963

5. Robert S. Knox
Theory of Excitons, 1963

6. S. Amelinckx
The Direct Observation of Dislocations, *in preparation*

ACADEMIC PRESS • New York and London • 1963

THEORY OF EXCITONS

ROBERT S. KNOX

Department of Physics and Astronomy
University of Rochester
Rochester, New York

ACADEMIC PRESS • New York and London • 1963

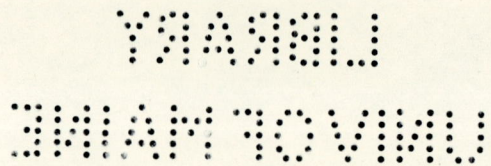

COPYRIGHT © 1963, BY ACADEMIC PRESS INC.

ALL RIGHTS RESERVED

NO PART OF THIS BOOK MAY BE REPRODUCED IN ANY FORM
BY PHOTOSTAT, MICROFILM, OR ANY OTHER MEANS,
WITHOUT WRITTEN PERMISSION FROM THE PUBLISHERS.

ACADEMIC PRESS INC.
111 FIFTH AVENUE
NEW YORK 3, N. Y.

United Kingdom Edition
Published by
ACADEMIC PRESS INC. (LONDON) LTD.
BERKELEY SQUARE HOUSE, LONDON W. 1

Library of Congress Catalog Card Number LC 63-22334

PRINTED IN THE UNITED STATES OF AMERICA

Preface

Exciton theory developed rather sporadically for about twenty years after Frenkel's fundamental work in 1931, and then experienced a sharp rise in the 1950's. This increased activity can be traced directly to the increased vigor with which optical properties of solids, particularly semiconductors, were studied by experimentalists during these years. At the present, a plateau seems to have been reached, which can be accounted for perhaps by the fact that the internal structure of the exciton and its most elementary dynamics are fairly well understood, at least in principle and in many cases in practice; theorists now await decisive experiments dealing with the energy transport and statistical behavior of the exciton. The plateau presents an ideal opportunity for a fairly comprehensive review of our present knowledge, and an attempt at such a review is presented herein.

The exciton usually "appears" as a characteristic feature in an absorption spectrum of a crystal, but does so in such a variety of ways that vastly different exciton languages have built up in different areas of solid-state and chemical physics. In molecular crystals, the exciton may be held responsible for certain splittings of molecular lines, while in semiconductors it may change the shape of a fundamental absorption edge as predicted by band theory and even add resolvable structure, the best-known example of which is the "hydrogenic" spectrum of Cu_2O. The exciton is responsible for the intense doublet lines discovered in the alkali halides four years before Frenkel's papers, and even enters into discussions of normal and superconducting metals. It has been hoped, in the preparation of this article, to present exciton theory in a manner in which an emphasis is placed on similarities, rather than differences, among excitons in different solids.

The author acknowledges with pleasure the helpful conversations he has had with many patient colleagues during the preparation of this article, in particular those with Professors K. Teegarden and A. Gold. Critical readings of the manuscript at various stages by Professors A. Gold and D. L. Dexter were greatly appreciated, as were translations of certain Russian articles generously provided by

Dr. C. C. Klick of the U. S. Naval Research Laboratory. The Air Force Office of Scientific Research supported a part of the work. Thanks are due to various workers who have granted permission to reproduce or adapt drawings. Finally, the author wishes to record his indebtedness to his wife, Mirta, whose constant encouragement made this work possible.

R. S. KNOX

The University of Rochester
Rochester, New York
September, 1963

Contents

Preface v

I. Introduction 1
   1. Survey 1

II. The Electronic Structure 7
   2. General Case: Use of the One-Electron Approximation . . . 7
   3. Frenkel's Case: Tight Binding 21
   4. Wannier's Case: Weak Binding 37
   5. Intermediate Cases 59
   6. Effects of Static External Fields 74
   7. Special Topics 87

III. Absorption and Dispersion of Light by Nonmetallic Solids . . . 103
   8. Classical Theory of the Optical Effects 103
   9. Semiclassical Theory of Optical Absorption 112
   10. Dynamical Effects of Phonons 136
   11. Anomalous Waves and spatial Dispersion 163

IV. Transport Phenomena and Related Topics 169
   12. Theory of Exciton Transport Phenomena 169
   13. Radiative Decay 188

V. Summary 191

AUTHOR INDEX 195

SUBJECT INDEX 201

I. Introduction

1. SURVEY

The theory of excitons was formulated thirty years ago by Frenkel, Peierls, and Wannier.[1-5] Their papers dealt with fundamental questions such as "What is the mechanism by which pure insulating solids absorb visible or ultraviolet light at a given wavelength? How do they dispose of the energy they thereby acquire?" Questions of this sort have been answered satisfactorily in the case of atoms and simple molecules, but in the case of solids, experimental facts immediately overwhelm any simple quatitative theory. For example, as shown in Figs. 1 and 2, in contrast with the well-known characteristics of atomic spectra, absorption lines of widely varying half-widths are observed in pure solids (from 0.005 ev in Cu_2O at low temperatures to 0.5 ev in some alkali halides at room temperature); emission subsequent to absorption is absent in a large percentage of cases. Unique polarization and magnetic field dependences of the absorption spectrum are present in certain solids. The frequent absence of emission, along with theoretical speculation on the fate of the absorbed energy, has inspired numerous experiments on crystals with controlled defect content, these defects being used as probes to capture the excitation energy.

One might hope to regard the absorption process in a solid simply in terms of the absorption of a dense gas. Broad lines and sometimes the absence of re-emission are seen in the case of impurity absorption in solids; in this case, a highly perturbed atom or molecular complex is essentially being observed. The case of the pure solid differs in one very important way which can be illustrated by analogy with phonon theory. It is known that the characterization of vibrational modes of

[1] J. Frenkel, *Phys. Rev.* **37**, 17 (1931).
[2] J. Frenkel, *Phys. Rev.* **37**, 1276 (1931).
[3] R. E. Peierls, *Ann. Physik* [5] **13**, 905 (1932).
[4] J. Frenkel, *Physik. Z. Sowjetunion* **9**, 158 (1936).
[5] G. H. Wannier, *Phys. Rev.* **52**, 191 (1937).

a crystal in terms of amplitudes of individual atomic displacements is not appropriate. The translational symmetry of the lattice demands that the true normal modes be linear combinations of these displacements, each mode belonging to a wave vector **f** in the reciprocal lattice. Similarly, one cannot expect to characterize an excited elec-

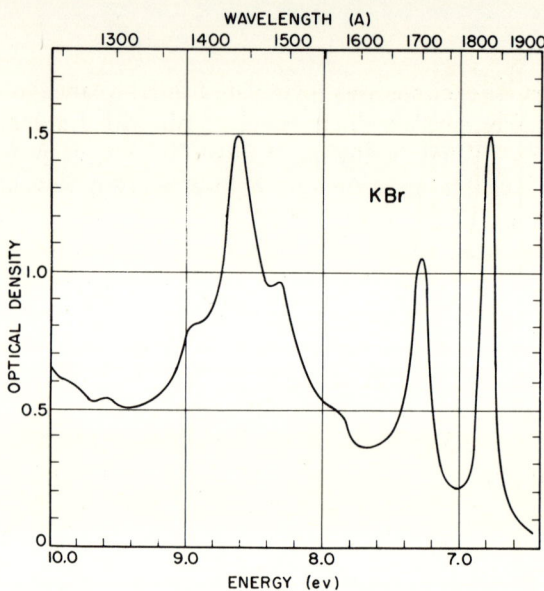

FIG. 1. Optical absorption spectrum of a thin film of KBr at liquid nitrogen temperature. Excitons are presumably created during absorption at 6.8 and 7.7 ev, and the fairly large widths are due to strong interactions of the excitons with imperfections such as phonons and surfaces. The optical density is proportional to the absorption coefficient, which is of the order of 10^6 cm^{-1} at the peaks. [After J. E. Eby, K. J. Teegarden, and D. B. Dutton, *Phys. Rev.* **116**, 1099 (1959).]

tronic state of a perfect crystal by localized states of electronic excitation. Each such state is equivalent to, or "in resonance with," excited states localized at other points in the lattice. True stationary excited electronic states must be linear combinations of these localized excitations, each belonging to a wave vector **K** in the reciprocal lattice. Frenkel called these running states of excitation "excitons," envisioning the construction of wave packets representing "particles" of excitation.

The wave-like nature of the exciton is a general consequence of the translational symmetry of the crystal and does not depend on the nature of the "localized excitation." In fact, the notion of a particular localized excitation, such as an excited atomic state, on which the above analogy is based, is not always useful. Wannier showed that an

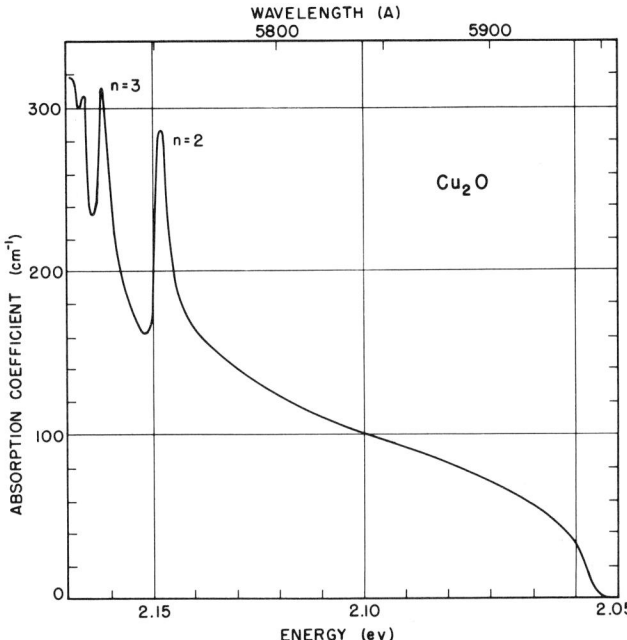

FIG. 2. Optical absorption spectrum of a crystal of Cu_2O at liquid helium temperature. Excitons are associated with the series of lines beginning at 2.149 ev, as well as with the continuous background. The line widths are narrower than those in Fig. 1 because there are fewer phonons and a smaller likelihood of surface effects, and smaller coupling constants involved in the interactions. [After P. W. Baumeister, *Phys. Rev.* **121**, 359 (1961).]

exciton may be viewed alternatively as a conduction-band electron and a valence-band hole, bound together but possibly with a considerable separation, the pair traveling through the crystal in a state of total wave vector **K**. These two pictures are not as unrelated as they might at first seem. Since an excited atom can be described essentially as an electron bound closely to an ion by the Coulomb interaction, the Frenkel and Wannier excitons differ physically merely in their "radii," i.e., in the degree of separation of the electron and hole, as shown schematically in Fig. 3. As will be seen, Wannier's model applies specifically to excitons of large radius, but an exciton of any intermediate radius can be imagined. The existence of such spatially extended bound states and the possibility of coherent migration of

Fig. 3. Illustrating the limiting cases of small and large exciton radii. In each case, there is internal motion of an electron around a hole and the pair moves through the lattice as a unit. Although equivalent in principle, these two kinds of excitons must be handled with quite different techniques because in the former case the electron "sees" both the hole and the details of the lattice potential, while in the latter it sees only the hole and an average lattice potential. The two cases are found in KBr and Cu_2O, respectively.

excitation make the exciton unique to solids and large regular molecules.

In spite of their long history, excitons have received vigorous attention only during the last few years. They are now of interest in every *type* of solid. The fundamental absorption spectra of ionic crystals, particularly the alkali halides, have been studied exhaustively ever since the pioneering work of Hilsch and Pohl.[6] These spectra are generally interpreted in terms of excitons having radii intermediate between Frenkel's and Wannier's.[6-11] Early Russian experimental[12] and theoretical[13] work on molecular crystals was the forerunner of

[6] R. Hilsch and R. W. Pohl, *Z. Physik* **48**, 384 (1928); **57**, 145 (1929); **59**, 812 (1930).
[7] K. L. Wolff and K. F. Herzfeld, in "Handbuch der Physik" (H. Geiger and K. Scheel, eds.), Vol. 20, p. 632. Springer, Berlin, 1928.
[8] A. von Hippel, *Z. Physik* **101**, 680 (1936).
[9] N. F. Mott, *Trans. Faraday Soc.* **34**, 500 (1938).
[10] D. L. Dexter, *Phys. Rev.* **83**, 435 (1951).
[11] A. W. Overhauser, *Phys. Rev.* **101**, 1702 (1956).
[12] I. W. Obreimov and A. F. Prikhot'ko, *Physik. Z. Sowjetunion* **9**, 48 (1936).
[13] A. S. Davydov, *Zh. Eksperim. i Teor. Fiz.* **18**, 210 (1948); see also Davydov's "Theory of Molecular Excitons" (transl. M. Kasha and M. Oppenheimer, Jr.). McGraw-Hill, New York, 1962.

extensive study in this area, recently reviewed by McClure[14] and by Wolf.[15] The possibility of exciton transfer of energy in these crystals is of considerable interest in connection with energy transfer in biological systems. Spectra of covalent solids (such as Ge, Si, and CdS) and ionic crystals with large dielectric constants (such as Cu_2O) have been studied thoroughly during the past ten years in work reviewed by Gross,[16] Ueta,[17] Nikitine,[18] and McLean.[19] These spectra are understood in great detail on the Wannier model, which has been given a more complete formulation by Dresselhaus[20] and Elliott.[21] Excitons in the solid rare gases, whose electronic structures resemble those of the alkali halides, were first studied experimentally by Schnepp and Dressler[22] and theoretically by the author[23] in a detailed treatment using the Frenkel model. Finally, in normal metals, ordinary excitons are expected to have too short a lifetime to be observable,[24] but the quasi-particle exciton appears as a possible state lying within the energy gap of a superconductor.[25]

By and large, exciton theory has developed along two distinct lines, which we shall call "exciton structure" and "exciton dynamics." The theory of exciton *structure* involves the study of the low-lying excited electronic states of the *perfect* crystal, its aim being to determine the detailed nature of the electron-hole wave functions and the corresponding excitation energies, including their dependence on the wave vector **K**. Exciton *dynamics* is defined as the study of interactions between excitons and other particles and fields; this is taken to include the exciton-photon interaction and the interaction of the exciton with imperfections such as phonons, dislocations, and point defects.

[14] D. S. McClure, *Solid State Phys.* **8**, 1 (1959).
[15] H. C. Wolf, *Solid State Phys.* **9**, 1 (1959).
[16] E. F. Gross, *Izv. Akad. Nauk S.S.S.R., Ser. Fiz.* **20**, 89 (1956); *Nuovo Cimento, Suppl.* **4**, 672 (1956); *Usp. Fiz. Nauk* **76**, 433 (1962) [*English transl.: Sov. Phys. (Uspekhi)* **5**, 195 (1962)].
[17] M. Ueta, *Progr. Theoret. Phys.* Suppl. no. 12, 40 (1959).
[18] S. Nikitine, *Phil. Mag.* **4**, 1 (1959); S. Nikitine, *in* "Progress in Semiconductors" (A. F. Gibson *et al.*, eds.), Vol. 6, p. 233. Wiley, New York, 1962.
[19] T. P. McLean, *in* "Progress in Semiconductors" (A. F. Gibson *et al.*, eds.), Vol. 5, p. 55. Wiley, New York, 1961.
[20] G. Dresselhaus, *Phys. Chem. Solids* **1**, 14 (1956).
[21] R. J. Elliott, *Phys. Rev.* **108**, 1384 (1957).
[22] O. Schnepp and K. Dressler, *J. Chem. Phys.* **33**, 49 (1960).
[23] R. S. Knox, *Phys. Chem. Solids* **9**, 238, 265 (1959).
[24] N. F. Mott, *Proc. Phys. Soc. (London)* **A62**, 416 (1949).
[25] A. Bardasis and J. R. Schrieffer, *Phys. Rev.* **121**, 1050 (1961).

Phonons, for example, scatter excitons, may lead to their self-trapping, and even create absorption bands where none are expected by breaking down selection rules. Much important work on the general problem of the exciton-phonon interaction has been done by Haken[26] and Toyozawa.[27] We hasten to point out that the exciton-phonon interaction cannot always be ignored or treated as a small perturbation on the energy levels of a system. For example, when the hole and electron are well separated, they may polarize the lattice (as well as the electronic system) individually, the result being a polaron bound to a hole polaron. Regardless of this fact, a vast literature exists on the exciton in perfect (and, in particular, static) crystals.

Part II deals with the exciton in a perfect crystal. After defining an idealized model and completing certain general computations in Section 2, we discuss the electronic structure of the Frenkel exciton (Section 3), that of the Wannier exciton (Section 4), and that of the exciton in crystals in which it is difficult to apply either theory with impunity (Section 5). Effects of external fields are treated in Section 6, and a brief discussion of statistical and many-body theory is included as Section 7. Part III deals with the dynamics of the absorption of light, beginning with a review of the classical theory (Section 8). Section 9 is devoted to the derivation of selection rules for direct absorption processes, based on the exciton-photon interaction, and in Section 10 the various important effects of phonons, such as line shifting, line broadening, and violation of momentum selection rules, are introduced. Section 11 briefly reviews recent Russian work on "anomalous waves." Finally, the "fate of the exciton" is studied in Part IV, Sections 12 and 13, in which transport phenomena and related subjects such as luminescence are discussed.

---

[26] H. Haken, *Fortschr. Physik* **6**, 271 (1958).
[27] Y. Toyozawa, *Progr. Theoret. Phys. (Kyoto)* **20**, 53 (1958).

II. The Electronic Structure

2. GENERAL CASE: USE OF THE ONE-ELECTRON APPROXIMATION[5,28]

a. An Idealized Model of an Insulator

It is possible to demonstrate most of the features of the electronic structure of the exciton on a very simple idealized model of an insulator which will, in fact, serve as the basis of most of the material discussed in this chapter, since it is readily generalized to embrace finer details encountered in treatments of specific solids. We consider a large number of atoms or ions arranged on a regular lattice whose unit cells are constructed on a Bravais lattice defined by the set of position vectors

$$\mathbf{R} = n_1 \mathbf{a}_1 + n_2 \mathbf{a}_2 + n_3 \mathbf{a}_3$$

in which the n_i run over all integers and the $\mathbf{a}_i$ are linearly independent basic translations. We will normally fix our attention on a large parallelepiped containing N unit cells, and unless otherwise indicated, Born–von Karman boundary conditions are imposed on the system. The coordinates of the electrons will be denoted by $\mathbf{r}_i$ and those of the nuclei by $\mathbf{X}_I$. In the absence of external fields, the total Hamiltonian of this system is

$$\mathcal{H} = -\sum_I \frac{\hbar^2 \nabla_I^2}{2M_I} - \sum_i \frac{\hbar^2 \nabla_i^2}{2m} + \sum_{I<J} \frac{z_I z_J e^2}{|\mathbf{X}_I - \mathbf{X}_J|}$$

$$- \sum_I \sum_i \frac{z_I e^2}{|\mathbf{X}_I - \mathbf{r}_i|} + \sum_{i<j} \frac{e^2}{|\mathbf{r}_i - \mathbf{r}_j|} + H_S \quad (2.1)$$

where the various terms are, respectively, the kinetic energy of the nuclei (∇_I is the gradient with respect to $\mathbf{X}_I$), the kinetic energy of the electrons, the potential energy of the nuclei interacting with each other, the potential energy of the electrons interacting with the nuclei,

[28] J. C. Slater and W. Shockley, *Phys. Rev.* **50**, 705 (1936).

the potential energy of the electrons interacting with each other, and the spin-dependent electronic energy H_S in which the usual spin-orbit interaction is assumed to be of greatest importance. We wish to find the eigenstates and eigenvalues of $\mathcal{H}$, which can hardly be started without the use of two well-known approximations.

At first, and throughout Chapter II, it will be assumed that the nuclei are at rest. The first term of Eq. (2.1) is thus effectively zero and the nuclear position vectors occupy equilibrium positions $\mathbf{R}_I$ on the perfect lattice. This implies separability of the electronic and nuclear wave functions; the electronic and vibrational excitations are to be treated separately at first and their interaction taken into account afterward. The Hamiltonian of the static crystal is thus

$$\mathcal{H}_0 = -\sum_i \frac{\hbar^2 \nabla_i^2}{2m} + \sum_i U(\mathbf{r}_i) + \sum_{i<j} \frac{e^2}{|\mathbf{r}_i - \mathbf{r}_j|} + H_S \qquad (2.2)$$

where $\Sigma_i\, U(\mathbf{r}_i)$ stands for the third and fourth terms of Eq. (2.1) with all $\mathbf{X}_I$ replaced by $\mathbf{R}_I$.

A second major assumption is that of the validity of the one-electron approximation. The solutions of

$$\mathcal{H}_0 \varphi(\mathbf{r}_1, \mathbf{r}_2, \cdots) = E \varphi(\mathbf{r}_1, \mathbf{r}_2, \cdots) \qquad (2.3)$$

are taken to be separable into antisymmetrized products of orthonormal one-electron functions,

$$\Phi = \mathcal{A} \varphi_1(\mathbf{r}_1)\, \varphi_2(\mathbf{r}_2) \cdots \qquad (2.4)$$

where $\mathcal{A} = (n!)^{-1/2} \Sigma_P (-1)^{p_P} P$, n is the number of electrons, P one of the $n!$ permutations of n objects, p_P its parity, and the sum runs over all $n!$ permutations. Only in the case of systems containing very few electrons has it been possible to avoid the one-electron approximation, and there appears to be little chance of doing so in solid state problems.

The computation of the ground-state energy E_0 for given solids, particularly metals, has been studied at length and reviewed frequently in recent years.[29] Even with knowledge of the best set of

---

[29] See, e.g., D. J. Thouless, "Quantum Mechanics of Many-Body Systems." Academic Press, New York, 1961; D. Pines, ed. "The Many-Body Problem." Benjamin, New York, 1961.

2. GENERAL CASE: USE OF THE ONE-ELECTRON APPROXIMATION

solutions of the form (2.4), i.e., the Hartree-Fock solutions, the true ground-state energy is not known because of the one-electron approximation, and some form of perturbation theory must be used to mix higher states built from one-electron functions into Eq. (2.4). The correction to the Hartree-Fock energy obtained in this or any other way is known as the correlation energy. We will be concerned with the energy differences involved in one-electron transitions, and an important "correlation energy" (the electron-hole interaction) appears even in first order. Higher-order effects are manifested in the introduction of a dielectric constant into this electron-hole interaction, as discussed in Sections 4 and 7.

The idealized model mentioned earlier has the following features. We visualize a monatomic crystal with no basis in which each constituent atom is assumed to contribute a core and two valence electrons. The core includes the nucleus and a charge cloud representing the remaining electrons; the potential produced by this cloud is best thought of as a self-consistent potential computed for the valence electrons, but in practice it must be chosen in some arbitrary though reasonable way. In this model, $\mathscr{H}_0$ is reinterpreted quite simply: the first term is the kinetic energy of the $2N$ valence electrons, the third their mutual Coulomb interaction, and the fourth their spin-orbit interaction. The second term now contains the core-valence electron and core-core interactions (previously we read "nucleus" for "core" and "electron" for "valence electron"). The ground electronic state in zeroth order is assumed to be a singlet in which the $2N$ valence electrons occupy N pairs of one-electron states, i.e.,

$$\Phi_0 = \mathscr{A} \varphi_{1\alpha}(\mathbf{r}_1) \, \varphi_{1\beta}(\mathbf{r}_2) \, \varphi_{2\alpha}(\mathbf{r}_3) \cdots \varphi_{N\beta}(\mathbf{r}_{2N}) \tag{2.5}$$

where α and β are spin quantum numbers ($m_s = \pm \frac{1}{2}$, respectively), and in which the coordinates $\mathbf{r}_i$ are taken to include a spin coordinate.[30]

Thus $\qquad \varphi_{i\alpha} \equiv \varphi_i(x, y, z) \, \chi_{1/2}(\sigma).$

Provided that the functions φ actually form an orthonormal set, Φ_0 is a normalized function and the first-order ground-state energy

---

[30] This convention, which applies only to arguments of one-electron functions, will be used hereafter whenever a spin quantum number appears in the designation of the function. Furthermore, $\int d\tau$ will imply summation over spin coordinates as well as integration over space coordinates, which will be written $\int d\mathbf{r}$.

is readily calculated in terms of matrix elements involving one-electron functions. The result is

$$E_0^{(1)} = (\Phi_0, \mathscr{H}_0 \Phi_0) = \sum_i \sum_\sigma \int \varphi_{i\sigma}(\mathbf{r})^* \left(-\frac{\hbar^2 \nabla^2}{2m} + U(\mathbf{r})\right) \varphi_{i\sigma}(\mathbf{r}) \, d\tau$$

$$+ \tfrac{1}{2} \sum_i \sum_\sigma \sum_j \sum_{\sigma'} \left(\iint \varphi_{i\sigma}(\mathbf{r}_1)^* \varphi_{j\sigma'}(\mathbf{r}_2)^* g_{12} \varphi_{i\sigma}(\mathbf{r}_1) \varphi_{j\sigma'}(\mathbf{r}_2) \, d\tau_1 \, d\tau_2 \right.$$

$$\left. - \iint \varphi_{i\sigma}(\mathbf{r}_1)^* \varphi_{j\sigma'}(\mathbf{r}_2)^* g_{12} \varphi_{j\sigma'}(\mathbf{r}_1) \varphi_{i\sigma}(\mathbf{r}_2) \, d\tau_1 \, d\tau_2 \right). \quad (2.6)$$

In this equation σ and σ' are summed from $-\tfrac{1}{2}$ to $+\tfrac{1}{2}$ and $g_{12} \equiv e^2/|\mathbf{r}_1 - \mathbf{r}_2|$. The second (exchange) g_{12} interaction vanishes, as usual, when $\sigma \neq \sigma'$.

The specification of the functions φ_i has thus far been avoided. As hinted earlier, they are best regarded as Hartree-Fock solutions of the entire system, but these solutions are unattainable in practice and one of two standard approximations can be used. On the one hand, the φ_i could be taken to be Bloch functions $\psi_{n\mathbf{k}}$, where n is a band index and $\mathbf{k} \, (= \mathbf{k}_1, \cdots, \mathbf{k}_N)$ ranges over the first BZ (Brillouin zone) of the lattice. On the other hand, according to tight binding theory, these functions could be taken as atomic functions $\phi_{l\mathbf{R}}$, where l stands for a set of atomic quantum numbers and $\mathbf{R} \, (= \mathbf{R}_1, \cdots, \mathbf{R}_N)$ is one of the N lattice sites at which $\phi_{l\mathbf{R}}$ is centered. In the former case the $\psi_{n\mathbf{k}}$ are solutions of the one-electron equation

$$\left[-\frac{\hbar^2 \nabla^2}{2m} + V(\mathbf{r})\right] \psi_{n\mathbf{k}} = \epsilon_{n\mathbf{k}} \psi_{n\mathbf{k}} \quad (2.7)$$

where $V(\mathbf{r})$ is an average, preferably self-consistent potential. Equation (2.7) can be used to simplify Eq. (2.6) somewhat, the result being

$$E_0^{(1)} = 2 \sum_{\mathbf{k}} \left[\epsilon_{m\mathbf{k}} + \int \psi_{m\mathbf{k}}(\mathbf{r})^* (U - V) \psi_{m\mathbf{k}}(\mathbf{r}) \, d\mathbf{r} \right]$$

$$+ \tfrac{1}{2} \sum_{\mathbf{k}} \sum_\sigma \sum_{\mathbf{k}'} \sum_{\sigma'} \left(\int \psi_{m\mathbf{k}\sigma}(\mathbf{r}_1)^* \psi_{m\mathbf{k}'\sigma'}(\mathbf{r}_2)^* g_{12} \psi_{m\mathbf{k}\sigma}(\mathbf{r}_1) \psi_{m\mathbf{k}'\sigma'}(\mathbf{r}_2) \, d\tau_1 \, d\tau_2 \right.$$

$$\left. - \int \psi_{m\mathbf{k}\sigma}(\mathbf{r}_1)^* \psi_{m\mathbf{k}'\sigma'}(\mathbf{r}_2)^* g_{12} \psi_{m\mathbf{k}'\sigma'}(\mathbf{r}_1) \psi_{m\mathbf{k}\sigma}(\mathbf{r}_2) \, d\tau_1 \, d\tau_2 \right). \quad (2.8)$$

This may be compared with the energy $2 \Sigma_{\mathbf{k}} \epsilon_{m\mathbf{k}}$ given by the simple Bloch scheme.

2. GENERAL CASE: USE OF THE ONE-ELECTRON APPROXIMATION

In the tight-binding case, the functions $\phi_{l\mathbf{R}}$ are found as solutions of the individual atomic Hamiltonians and an expression for $E_0^{(1)}$ similar to Eq. (2.8) results. Provided the $\phi_{l\mathbf{R}}$ satisfy

$$\left[-\frac{\hbar^2 \nabla^2}{2m} + u(\mathbf{r} - \mathbf{R})\right] \phi_{l\mathbf{R}} = \epsilon_l \phi_{l\mathbf{R}} \qquad (2.9)$$

where $u(\mathbf{r} - \mathbf{R})$ is the potential of a single core centered at $\mathbf{R}$, we have

$$E_0^{(1)} = 2 \sum_{\mathbf{R}} \left[\epsilon_l + \int \phi_{l\mathbf{R}}^* \left(U - \sum_{\mathbf{R}'} u(\mathbf{r} - \mathbf{R}')\right) \phi_{l\mathbf{R}} \, d\mathbf{r}\right]$$
$$+ \frac{1}{2} \sum_{\mathbf{R}} \sum_{\sigma} \sum_{\mathbf{R}'} \sum_{\sigma'} \left(\int \phi_{l\mathbf{R}\sigma}(\mathbf{r}_1)^* \phi_{l\mathbf{R}'\sigma'}(\mathbf{r}_2)^* g_{12} \phi_{l\mathbf{R}\sigma}(\mathbf{r}_1) \phi_{l\mathbf{R}'\sigma'}(\mathbf{r}_2) \, d\tau_1 \, d\tau_2\right.$$
$$\left.- \int \phi_{l\mathbf{R}\sigma}(\mathbf{r}_1)^* \phi_{l\mathbf{R}'\sigma'}(\mathbf{r}_2)^* g_{12} \phi_{l\mathbf{R}'\sigma'}(\mathbf{r}_1) \phi_{l\mathbf{R}\sigma}(\mathbf{r}_2) \, d\tau_1 \, d\tau_2\right). \qquad (2.10)$$

Here the lowest "atomic" state l is chosen in computing the goundstate energy, and in Eq. (2.8) the band index m refers to the filled valence band.

Equations (2.9) and (2.10) have been presented in a form best suited to comparison with Eqs. (2.7) and (2.8). The eigenvalue problem (2.9) ignores the interaction between the two valence electrons of the atom and this interaction reappears somewhat unnaturally as a correction to the atomic eigenvalues in the second and third terms of Eq. (2.10). In his applications of the tight-binding theory to the problem of cohesive energies of the alkali halides, Löwdin[31] used instead of Eq. (2.9) the atomic Hartree-Fock equations. The intra-atomic interactions are then entirely included in ϵ_l and terms with $\mathbf{R} = \mathbf{R}'$ in the second and third terms of Eq. (2.10) are absent. Another extremely important assumption made in the simplified tight-binding scheme presented here is the assumption that atomic functions centered at different sites are orthogonal. Since they hardly ever are, in practice, special measures such as Löwdin's symmetric orthogonalization must be taken to validate Eq. (2.10). An excellent discussion of this procedure is found in the article by Reitz.[32]

The two alternative descriptions of one-electron states presented above differ in that the former, $\psi_{n\mathbf{k}} = e^{i\mathbf{k}\cdot\mathbf{r}} u_{n\mathbf{k}}(\mathbf{r})$, are running waves

---

[31] P. O. Löwdin, *Arkiv Mat. Astron. Fysik* **35**, No. 9 (1948).
[32] J. R. Reitz, *Solid State Phys.* **1**, 1 (1955).

distributed rather uniformly over the entire crystal, while the latter, $\phi_{l\mathbf{R}}$, are by construction localized. It is often pointed out that in a crystal such as the one described by our model, i.e., one consisting of atoms in closed shells, the ground state is identical in the two schemes. This holds in the following formal sense: if from the set of localized states one constructs Bloch states

$$\psi_{l\mathbf{k}} = N^{-1/2} \sum_{\mathbf{R}} e^{i\mathbf{k}\cdot\mathbf{R}} \phi_{l\mathbf{R}}(\mathbf{r}) \qquad (2.11)$$

then the determinantal functions

$$\mathscr{A} \psi_{l\mathbf{k}_1\alpha}(\mathbf{r}_1) \psi_{l\mathbf{k}_1\beta}(\mathbf{r}_2) \cdots \psi_{l\mathbf{k}_N\beta}(\mathbf{r}_{2N}) \qquad (2.12)$$

and

$$\mathscr{A} \phi_{l\mathbf{R}_1\alpha}(\mathbf{r}_1) \phi_{l\mathbf{R}_1\beta}(\mathbf{r}_2) \cdots \phi_{l\mathbf{R}_N\beta}(\mathbf{r}_{2N}) \qquad (2.13)$$

are identically equal. [This can be seen at once by writing Eqs. (2.12) and (2.13) as determinants, and noting that the elements form matrices related by simple multiplication by the unitary matrix $U(U_{pq} = N^{-1/2} e^{i\mathbf{k}_p \cdot \mathbf{R}_q})$ whose determinant is unity. Thus the theorem is valid for states describable by single determinants, true for the zero-order description of closed-shell systems.] It follows that $E_0^{(1)}$ is the same in both representations (2.12) and (2.13). The tight-binding Bloch states (2.11) are useful only under certain conditions such as small overlap of atomic functions, and do not provide a good description of most semiconductor valence bands. Wannier pointed out, however, that the inverse of Eq. (2.11) can be used to construct localized functions if one wishes to work with Bloch functions obtained by methods more accurate than those obtained from tight binding theory. He wrote this inverse relationship as

$$a_{n\mathbf{R}} = N^{-1/2} \sum_{\mathbf{k}} e^{-i\mathbf{k}\cdot\mathbf{R}} \psi_{n\mathbf{k}}(\mathbf{r}) \qquad (2.14)$$

where we retain his original a notation to emphasize that these functions, localized about $\mathbf{R}$, are not necessarily atomic functions $\phi_{n\mathbf{R}}$, and in fact never will be unless the $\psi_{n\mathbf{k}}$ on the right-hand side happen to be given by Eq. (2.11). The symmetry properties and degree of localization of these Wannier functions have been studied by Slater,[33]

[33] J. C. Slater, *Phys. Rev.* **76**, 1592 (1949); **87**, 807 (1952).

2. GENERAL CASE: USE OF THE ONE-ELECTRON APPROXIMATION

who emphasized the importance of Wannier's approach, by Koster,[34] and by Kohn.[35] By our earlier discussion we know that the states

$$\mathscr{A}\psi_{nk_1\alpha}(\mathbf{r}_1)\,\psi_{nk_1\beta}(\mathbf{r}_2)\cdots\psi_{nk_N\beta}(\mathbf{r}_{2N}) \quad (2.15)$$

and

$$\mathscr{A}a_{n\mathbf{R}_1\alpha}(\mathbf{r}_1)\,a_{n\mathbf{R}_1\beta}(\mathbf{r}_2)\cdots a_{n\mathbf{R}_N\beta}(\mathbf{r}_{2N}) \quad (2.16)$$

are identical, and that $E_0^{(1)}$ is therefore the same computed in either the ψ_{nk} or $a_{n\mathbf{R}}$ representation.

b. Excited States in the Idealized Model

In Section 2a we have discussed the description of the ground state of an idealized crystal in terms of localized and Bloch states and have emphasized the equivalence of the two descriptions when-

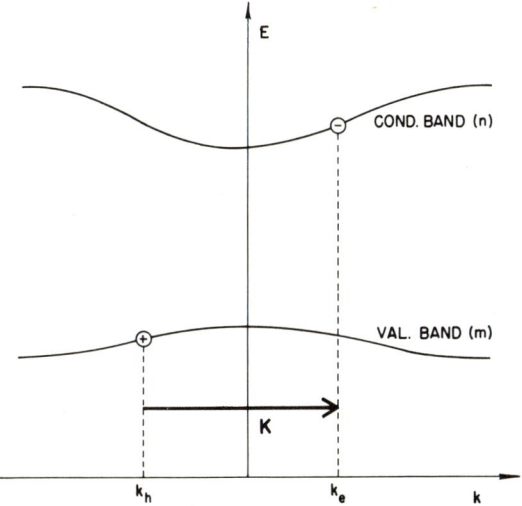

FIG. 4. The wave vectors involved in zero-order states of excitation in the Bloch scheme. The ordinate is energy, the abscissa wave vector. The lower curve represents an almost filled valence band with a hole, and the upper curve an almost empty conduction band with one electron.

---

[34] G. F. Koster, *Phys. Rev.* **89**, 67 (1953).
[35] W. Kohn, *Phys. Rev.* **115**, 809 (1959).

ever the localized and Bloch one-electron functions are uniquely related to one another. In the case of excited states, *neither* description is adequate, as will be seen below. First, to complete the specification of the idealized model, we assume that the lowest zero-order excited states lie appreciably above the ground state; for example, on the band picture, there is assumed to be a definite energy gap in the eigenvalues $\epsilon_{n\mathbf{k}}$. Specifically, we will consider a conduction band (labeled n in Fig. 4) near whose minimum any additional electrons would reside if added to the idealized crystal.

An attempt will first be made to describe the elementary excitations in terms of removal of a valence electron from a Bloch state to another Bloch state in the conduction band. This is reasonable on two counts. Koopmans's theorem,[36] crude as it is in this application, suggests that these excitations will be involved in constructing the lowest excited state, since they require the least energy in the approximation which neglects all correlation. Also, any strong optical excitations will involve only one electron. Beginning with the ground state (2.15), we obtain therefore $(2N)^2$ possible zero-order excited states of the form

$$\Phi_{mn}^{\sigma\sigma'}(\mathbf{k}_h, \mathbf{k}_e) = \mathscr{A}\psi_{m\mathbf{k}_1\alpha}\psi_{m\mathbf{k}_1\beta}\cdots\psi_{m\mathbf{k}_h\sigma}\psi_{n\mathbf{k}_e\sigma'}\cdots\psi_{m\mathbf{k}_N\beta}, \qquad (2.17)$$

a state in which a valence band electron of spin $-\sigma$ and momentum $\mathbf{k}_h$ has been removed to a state in the conduction band (n) with spin σ' and momentum $\mathbf{k}_e$. This state is shown schematically in Fig. 4. The total momentum of the system is now $\mathbf{k}_e - \mathbf{k}_h \equiv \mathbf{K}$ and the eigenvalue of the component of the total spin along the axis of quantization is $\frac{1}{2}(\sigma + \sigma')\hbar$. Eigenfunctions of definite multiplicity can easily be constructed from states of the form (2.17); for example, a state $\Phi_{mn}^{\alpha\alpha}$ is definitely a member of a triplet, while one singlet and one triplet state are obtained from appropriate linear combinations of $\Phi_{mn}^{\alpha\beta}$ and $\Phi_{mn}^{\beta\alpha}$. On the simple Bloch scheme the energy of this state relative to that of the ground state is $\epsilon_{n\mathbf{k}_e} - \epsilon_{m\mathbf{k}_h}$. The wave vector $\mathbf{K} = \mathbf{k}_e - \mathbf{k}_h$ is significant in that it characterizes the way in which the total wave function transforms under the translational symmetry operations of the crystal. Mixing is expected among only those zero-order functions having the same reduced value[37] of $\mathbf{k}_e - \mathbf{k}_h$. It is a relatively straight-

---

[36] See, e.g., F. Seitz, "Modern Theory of Solids," McGraw-Hill, New York, 1940. Sect. 67.

[37] The same value to within an additive multiple of 2π times a reciprocal lattice vector.

2. GENERAL CASE: USE OF THE ONE-ELECTRON APPROXIMATION

forward matter to compute matrix elements of $\mathcal{H}_0$ [Eq. (2.2)] connecting states of the form (2.17), and the result is

$$(mn\mathbf{k}_e\mathbf{k}_h \mid \mathcal{H}_0 \mid mn\mathbf{k}_{e'}\mathbf{k}_{h'}) = \delta_{\mathbf{k}_h\mathbf{k}_{h'}}\delta_{\mathbf{k}_e\mathbf{k}_{e'}}[E_0^{(1)} + W_n(\mathbf{k}_e) - W_m(\mathbf{k}_h)]$$
$$+ \delta_{\mathbf{k}_e-\mathbf{k}_h,\mathbf{k}_{e'}-\mathbf{k}_{h'}}[2\delta_M(n\mathbf{k}_e m\mathbf{k}_{h'} \mid g \mid m\mathbf{k}_h n\mathbf{k}_{e'}) - (n\mathbf{k}_e m\mathbf{k}_{h'} \mid g \mid n\mathbf{k}_{e'} m\mathbf{k}_h)]. \quad (2.18)$$

Spin indices have been omitted since this expression has been derived for true singlet states (taking $\delta_M = 1$) and pure triplets (taking $\delta_M = 0$). Each term has a fairly straightforward physical meaning. The first line is the energy of the zero-order excited state in an approximation similar to the one leading to the ground-state energy Eq. (2.8). $E_0^{(1)}$ is this ground-state energy, and $W_n(\mathbf{k})$ is the energy of an extra electron in conduction band n with wave vector $\mathbf{k}$, including its Coulomb and exchange interaction with the electrons present in valence band m:

$$W_n(\mathbf{k}) = \epsilon_n(\mathbf{k}) + (n\mathbf{k} \mid U - V \mid n\mathbf{k})$$
$$+ \sum_{\mathbf{k}'} [2(n\mathbf{k}m\mathbf{k}' \mid g \mid n\mathbf{k}m\mathbf{k}') - (n\mathbf{k}m\mathbf{k}' \mid g \mid m\mathbf{k}'n\mathbf{k})]. \quad (2.19)$$

Similarly, $W_m(\mathbf{k})$ is the energy of one valence electron:

$$W_m(\mathbf{k}) = \epsilon_m(\mathbf{k}) + (m\mathbf{k} \mid U - V \mid m\mathbf{k})$$
$$+ \sum_{\mathbf{k}'} [2(m\mathbf{k}m\mathbf{k}' \mid g \mid m\mathbf{k}m\mathbf{k}') - (m\mathbf{k}m\mathbf{k}' \mid g \mid m\mathbf{k}'m\mathbf{k})]. \quad (2.20)$$

Throughout, the usual convention for one- and two-electron integrals is being used, e.g.,

$$(m\mathbf{k}n\mathbf{k}' \mid g \mid m\mathbf{k}''n\mathbf{k}''') = \iint \psi_{m\mathbf{k}}(\mathbf{r}_1)^* \, \psi_{n\mathbf{k}'}(\mathbf{r}_2)^* \, g_{12} \, \psi_{m\mathbf{k}''}(\mathbf{r}_1) \, \psi_{n\mathbf{k}'''}(\mathbf{r}_2) \, d\mathbf{r}_1 \, d\mathbf{r}_2 \quad (2.21)$$

The second and third lines of Eq. (2.18) have simple interpretations only in the diagonal element ($\mathbf{k}_h = \mathbf{k}_h'$ and $\mathbf{k}_e = \mathbf{k}_e'$), where they are seen to be the exchange and Coulomb interactions, respectively, between the electron and hole. The exchange interaction disappears in the triplet state because the hole and electron have parallel spins. The spin-orbit interaction has been neglected here for simplicity. Instead of attempting to diagonalize $\mathcal{H}_0$ in this representation, let us examine the situation when viewed in terms of localized functions.

A state in which one electron of spin $-\sigma$ has been taken from a band m Wannier state in cell $\mathbf{R}_i$ and placed in cell $\mathbf{R}_j$ in a band n Wannier state with spin σ' is written

$$\Phi_{mn}^{\sigma\sigma'}(\mathbf{R}_i, \mathbf{R}_j) = \mathscr{A} a_{m\mathbf{R}_1\alpha} a_{m\mathbf{R}_1\beta} \cdots a_{m\mathbf{R}_i\sigma} a_{n\mathbf{R}_j\sigma'} \cdots a_{m\mathbf{R}_N\beta}. \tag{2.22}$$

This zero-order state is shown schematically in Fig. 5. The zero-

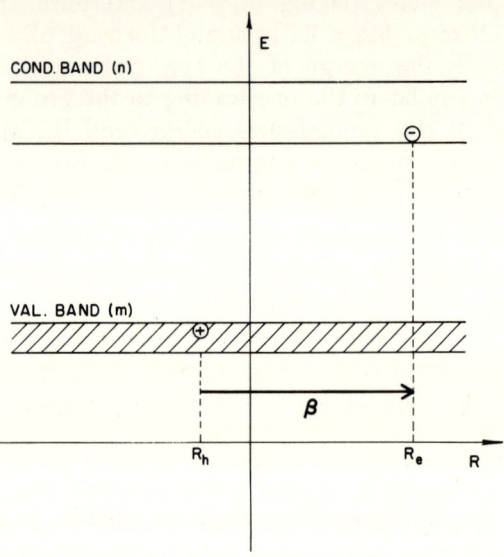

FIG. 5. The position vectors involved in zero-order states of excitation in a localized representation. The ordinate is energy, the abscissa position of lattice cells. The lower and upper bands are nearly full and nearly empty, respectively.

order states (2.17) and (2.22) are certainly not identical, in contrast with the situation in the ground state; they are, in fact, each other's Fourier transform. Explicitly,

$$\Phi_{mn}(\mathbf{k}_h, \mathbf{k}_e) = N^{-1} \sum_i \sum_j e^{i(\mathbf{k}_e'\cdot\mathbf{R}_j - \mathbf{k}_h'\cdot\mathbf{R}_i)} \Phi_{mn}(\mathbf{R}_i, \mathbf{R}_j). \tag{2.23}$$

To find the first-order excited eigenstates of the crystal using this localized representation, one must in principle diagonalize the matrix

$$(mn\mathbf{R}_i\mathbf{R}_j | \mathscr{H}_0 | mn\mathbf{R}_i'\mathbf{R}_j') \tag{2.24}$$

2. GENERAL CASE: USE OF THE ONE-ELECTRON APPROXIMATION

which has a complicated analytical form involving integrals of the form

$$(m\mathbf{R}n\mathbf{R}' \mid g \mid m\mathbf{R}''n\mathbf{R}''') = \iint a_{m\mathbf{R}}(\mathbf{r}_1)^* a_{n\mathbf{R}'}(\mathbf{r}_2)^* g_{12} \, a_{m\mathbf{R}''}(\mathbf{r}_1) \, a_{n\mathbf{R}'''}(\mathbf{r}_2) \, d\mathbf{r}_1 \, d\mathbf{r}_2 \tag{2.25}$$

and

$$W(n\mathbf{R}, n\mathbf{R}') = (n\mathbf{R} \mid -\hbar^2 \nabla^2/2m + U \mid n\mathbf{R}')$$

$$+ \sum_{\mathbf{R}''} [2(n\mathbf{R}m\mathbf{R}'' \mid g \mid n\mathbf{R}'m\mathbf{R}'') - (n\mathbf{R}m\mathbf{R}'' \mid g \mid m\mathbf{R}''n\mathbf{R}')] \, . \tag{2.26}$$

The first of these is a typical Coulomb integral and the second represents the interaction of the charge distribution $a_{n\mathbf{R}}^* a_{n\mathbf{R}'}$ with the rest of the crystal.

Although it is not necessary to write Eq. (2.24) out explicitly, it is important to note that translational symmetry simplifies it in the following way. If we write $\mathbf{R}_j = \mathbf{R}_i + \boldsymbol{\beta}$ and $\mathbf{R}_j' = \mathbf{R}_i' + \boldsymbol{\beta}'$, then Eq. (2.24) is a function only of $\mathbf{R}_i' - \mathbf{R}_i$, $\boldsymbol{\beta}$, and $\boldsymbol{\beta}'$. This fact is useful in making a transformation to the "exciton" representation discussed in the following.

Neither of the two representations discussed above provides a useful Hamiltonian matrix for the true excited state. On the one hand, a single **K** vector should be explicit in the wave function, since the total momentum of the electron-hole pair should be a good quantum number; this feature is provided only by the Bloch representation. On the other hand, we expect some explicit dependence of the matrix elements on $\boldsymbol{\beta}$, the electron-hole separation, as manifested in the extensive translational degeneracy in the localized representation. In view of this, Wannier suggested a third representation called the "exciton representation," related to the previous two by the simple unitary transformations

$$\Phi_{mn}(\mathbf{K}, \boldsymbol{\beta}) = N^{-1/2} \sum_{\mathbf{k}} e^{-i\boldsymbol{\beta}\cdot\mathbf{k}} \, \Phi_{mn}(\mathbf{k} - \mathbf{K}, \mathbf{k}) \tag{2.27}$$

$$= N^{-1/2} \sum_{\mathbf{R}} e^{i\mathbf{k}\cdot\mathbf{R}} \, \Phi_{mn}(\mathbf{R}, \mathbf{R} + \boldsymbol{\beta}) \, . \tag{2.28}$$

Provided the remarks on translational symmetry in the last paragraph are borne in mind, the matrix of $\mathscr{H}_0$ in the exciton representation is easily computed in terms of the interaction introduced earlier. The result is

$$(mn\mathbf{K}\beta \mid \mathscr{H}_0 \mid mn\mathbf{K}\beta') = \delta_{\beta\beta'} E_0^{(1)} + W(n\beta, n\beta') - e^{i\mathbf{K}\cdot(\beta-\beta')} W(m\mathbf{O}, m\beta - \beta')$$
$$+ 2\delta_M(n\beta m\mathbf{O} \mid g \mid m\mathbf{O}n\beta') - (n\beta m\mathbf{O} \mid g \mid n\beta' m\mathbf{O})$$
$$+ \sum_{\mathbf{R}\neq 0} e^{i\mathbf{K}\cdot\mathbf{R}}[2\delta_M(n\beta m\mathbf{R} \mid g \mid m\mathbf{O}n\mathbf{R}+\beta') - (n\beta m\mathbf{R} \mid g \mid n\mathbf{R}+\beta' m\mathbf{O})].$$
(2.29)

The notation of Eqs. (2.25) and (2.26) is used, with the additional use of symbols such as $n\mathbf{O}$, $n\beta$, and $n\mathbf{R}+\beta$, which stand for one-electron states from band n localized at the site arbitrarily chosen to lie at the origin, at β, and at $\mathbf{R}+\beta$, respectively. The diagonal elements of Eq. (2.29) show clearly the contributions of the electron and hole to the excitation energy and, in addition, their Coulomb and exchange interactions with each other. The term involving a lattice sum will be interpreted in Sections 3 and 4.

It will be convenient later to have transformed the second and third terms on the right band side of Eq. (2.29) into a form involving modified Bloch eigenvalues. Thus, using Eqs. (2.14), (2.19), and (2.26), we find

$$W(n\beta, n\beta') - e^{i\mathbf{K}\cdot(\beta-\beta')} W(m\mathbf{O}, m\beta - \beta')$$
$$= N^{-1} \sum_{\mathbf{k}} e^{i\mathbf{k}\cdot(\beta-\beta')} [W_n(\mathbf{k}) - W_m(\mathbf{k}-\mathbf{K})]$$
$$= e^{i\mathbf{K}\cdot(\beta-\beta')/2} N^{-1} \sum_{\mathbf{k}} e^{i\mathbf{k}\cdot(\beta-\beta')} [W_n(\mathbf{k}+\tfrac{1}{2}\mathbf{K}) - W_m(\mathbf{k}-\tfrac{1}{2}\mathbf{K})]. \quad (2.30)$$

The second equivalent form of the exciton representation wave function, Eq. (2.28), suggests quite clearly a picture of a hole at one lattice site and an electron at another at a distance β, the pair traveling in a state of total momentum $\mathbf{K}$. The case $\beta = 0$ corresponds formally to the Frenkel exciton, since $\Phi_{mn}(\mathbf{R}, \mathbf{R})$ is a state in which the hole and electron are at precisely the same lattice site, i.e., the atom at $\mathbf{R}$ is in an "excited state." As noted above, the expectation value of $\mathscr{H}_0$ in such a state is independent of $\mathbf{R}$ and an N-fold degeneracy exists. The linear combination (2.28) lifts this degeneracy. More generally, the expectation value of $\mathscr{H}_0$ in any state $\Phi_{mn}(\mathbf{R}, \mathbf{R}+\beta)$ is independent

2. GENERAL CASE: USE OF THE ONE-ELECTRON APPROXIMATION

of **R** and the function (2.28) lifts the resulting N-fold degeneracy. It remains, then, to find the linear combinations of states (2.28) with respect to the index β which complete the diagonalization of $\mathscr{H}_0$. [Under ideal conditions (Section 4), the probability in a given stationary state of the system of finding the electron at a position β with respect to the hole is given by a hydrogen-like envelope function with argument β.] These linear combinations will be written in general as

$$\Psi_{mn\nu\mathbf{K}} = \sum_{\beta} U_{mn\nu\mathbf{K}}(\beta)\, \Phi_{mn}(\mathbf{K}, \beta) \tag{2.31}$$

and the coefficients U are to be determined from the set of equations

$$\sum_{\beta'} (mn\mathbf{K}\beta \mid \mathscr{H}_0 \mid mn\mathbf{K}\beta')\, U_{mn\nu\mathbf{K}}(\beta') = E U_{mn\nu\mathbf{K}}(\beta) \tag{2.32}$$

subject to the usual secular equation for the eigenvalues E:

$$\det \mid (mn\mathbf{K}\beta \mid \mathscr{H}_0 \mid mn\mathbf{K}\beta') - E\delta_{\beta\beta'} \mid = 0. \tag{2.33}$$

The most serious deficiency of the formalism just completed is the restriction of the hole and electron to two *particular* bands, m and n. There exist possible states of higher energy in which the electron is in a higher conduction band or the hole is in a lower normally filled band. Formally, these states may be accounted for on the idealized model by including them in the zero-order set and seeking eigenstates of the much larger matrix

$$(mn\mathbf{K}\beta \mid \mathscr{H}_0 \mid m'n'\mathbf{K}\beta'). \tag{2.34}$$

Corrected energies for the zero-order states $\Psi_{mn\nu\mathbf{K}}$ might then be obtained through the use of perturbation theory. This is a difficult problem which must be handled by the techniques of many-body theory, and we will return to it in Section 7. The main effect on the two-band results is the introduction of a dielectric constant into the electron-hole interaction, since the inclusion of higher states allows polarization of the medium by these particles. Even though we will not use the matrix element (2.34) explicitly in such perturbation corrections, it will be important to know its value in cases involving degenerate or nearly degenerate bands (as, for example, in the p-like valence bands of alkali halides). Thus we make the following obvious generalization of Eq. (2.29); it is understood that $m \not\equiv n'$ and $m' \not\equiv n$:

$$(mn\mathbf{K}\boldsymbol{\beta} \mid \mathscr{H}_0 \mid m'n'\mathbf{K}\boldsymbol{\beta}') = \delta_{mm'}\delta_{nn'}\delta_{\boldsymbol{\beta}\boldsymbol{\beta}'}E_0^{(1)}$$
$$+ \delta_{mm'}W(n\boldsymbol{\beta}, n'\boldsymbol{\beta}') - \delta_{nn'}e^{i\mathbf{K}\cdot(\boldsymbol{\beta}-\boldsymbol{\beta}')}W(m\mathbf{O}, m'\boldsymbol{\beta}-\boldsymbol{\beta}')$$
$$+ 2\delta_M(n\boldsymbol{\beta}m'\mathbf{O} \mid g \mid m\mathbf{O}n'\boldsymbol{\beta}') - (n\boldsymbol{\beta}m'\mathbf{O} \mid g \mid n'\boldsymbol{\beta}'m\mathbf{O})$$
$$+ \sum_{\mathbf{R}} e^{i\mathbf{K}\cdot\mathbf{R}}[2\delta_M(n\boldsymbol{\beta}m'\mathbf{R} \mid g \mid m\mathbf{O}n'\mathbf{R}+\boldsymbol{\beta}') - (n\boldsymbol{\beta}m'\mathbf{R} \mid g \mid n'\mathbf{R}+\boldsymbol{\beta}'m\mathbf{O})].$$
(2.35)

The two-band formalism is adequate to handle crystals with a basis, although it may be convenient in certain cases to construct exciton states from hole states explicitly associated with one sublattice (e.g., the Cl⁻ sublattice of NaCl) and the electrons associated with the other (e.g., the Na⁺ sublattice of NaCl). This matter will be considered in Section 5.

The spin-orbit interaction will generally contribute two terms to the diagonal elements of $\mathscr{H}_0$ in each of the representations considered, one for the electron and one for the hole. The detailed form of the contributions depends on the symmetry of the spatial part of the electron and hole wave functions and on their relative spins, so it is not convenient to write a general formula. However, in an abstract form, the spin-orbit matrix element connecting the localized states $\Phi_{mn}^{\alpha\alpha}(\mathbf{R}, \mathbf{R}')$ and $\Phi_{mn}^{\alpha\alpha}(\mathbf{R}'', \mathbf{R}''')$ is

$$\delta_{\mathbf{R}\mathbf{R}''} \int a_{n\mathbf{R}'\alpha}(\mathbf{r})^* \, h a_{n\mathbf{R}'''\alpha}(\mathbf{r}) \, d\tau + \delta_{\mathbf{R}'\mathbf{R}'''} \int a_{m\mathbf{R}\alpha}(\mathbf{r})^* \, h a_{m\mathbf{R}''\alpha}(\mathbf{r}) \, d\tau. \quad (2.36)$$

[Recall that the spin of the hole is $-\frac{1}{2}$. In Eq. (2.36), we have used $H_S = \Sigma_i h(\mathbf{r}_i, \sigma_i)$.] Finally, it should be noted that the formalism of this section has been derived in the "LS coupling" limit in the sense that the total spin quantum number has been assumed good. Exciton states can just as easily be constructed from valence and conduction states whose space and spin parts have already been coupled in order to give them the correct symmetry behavior under the operations of the double crystal group. In this case, in order to be useful for detailed calculations, the matrix elements of $\mathscr{H}_0$ [Eqs. (2.18), (2.24), and (2.29)] need to be transformed to the "jj limit," but here again the result would depend on the specific symmetries of the states involved.

3. Frenkel's Case: Tight Binding

Historically, the theory of the electronic structure did not develop along the lines of Section 2. Originally Frenkel proposed[1] trial excited states constructed from ground and excited states of isolated atoms. Slater and Shockley[28] studied the transformation (2.23) from states involving Bloch pairs of localized functions, and finally Wannier[5] gave the connection between the two pictures. It is the purpose of this section to study the Frenkel model and its development.

a. Eigenvalues and Eigenfunctions

We shall first obtain Frenkel's states and eigenvalues by following what amounts to the original derivation, and then study them in more detail by applying the general results of Section 2. On a model similar to our idealized model but in which spin is ignored and only one valence electron is assigned to an atom,[38-41] the zero-order ground state is

$$\Phi_0 = \mathscr{A} \phi_{0\mathbf{R}_1}(\mathbf{r}_1) \phi_{0\mathbf{R}_2}(\mathbf{r}_2) \cdots \phi_{0\mathbf{R}_N}(\mathbf{r}_N). \tag{3.1}$$

Here $\phi_{0\mathbf{R}}(\mathbf{r})$ is the ground-state wave function of the atom located at site $\mathbf{R}$, and these N one-electron functions are assumed to be non-overlapping. An excited state can be obtained tentatively by the same construction if it is assumed that the atom at site $\mathbf{R}_i$ is in its lth excited state, $\phi_{l\mathbf{R}_i}(\mathbf{r})$, and this tentative state is written

$$\Phi_l(\mathbf{R}_i) = \mathscr{A} \phi_{0\mathbf{R}_1}(\mathbf{r}_1) \cdots \phi_{l\mathbf{R}_i}(\mathbf{r}_i) \cdots \phi_{0\mathbf{R}_N}(\mathbf{r}_N), \tag{3.2}$$

the same product as Eq. (3.1) with the exception that $\phi_{0\mathbf{R}_i}$ is replaced by $\phi_{l\mathbf{R}_i}$. It is clear from the translational symmetry of the crystal that the zero-order energies

$$E_{ii}^l = \int \Phi_l(\mathbf{R}_i)^* \mathscr{H}_0 \Phi_l(\mathbf{R}_i) \, d\mathbf{r}_1 \cdots d\mathbf{r}_N \tag{3.3}$$

are independent of $\mathbf{R}_i$, resulting in an N-fold degeneracy. Further-

---

[38] F. Seitz, "Modern Theory of Solids," p. 414. McGraw-Hill, New York, 1940.
[39] W. R. Heller and A. Marcus, *Phys. Rev.* **84**, 809 (1951).
[40] D. L. Dexter and W. R. Heller, *Phys. Rev.* **91**, 273 (1953).
[41] D. L. Dexter, *Phys. Rev.* **101**, 48 (1956).

more, the crystal Hamiltonian has nonvanishing matrix elements[42] between states (3.2) in which the excitation is at different sites:

$$E_{ii'}^l = \int \Phi_l(\mathbf{R}_i)^* \mathcal{H}_0 \, \Phi_l(\mathbf{R}_{i'}) \, d\mathbf{r}_1 \cdots d\mathbf{r}_N$$

$$= (l\mathbf{R}_i 0\mathbf{R}_{i'} \,|\, g \,|\, 0\mathbf{R}_i l\mathbf{R}_{i'}) - (l\mathbf{R}_i 0\mathbf{R}_{i'} \,|\, g \,|\, l\mathbf{R}_i 0\mathbf{R}_i). \quad (3.4)$$

It follows that the states (3.2) are not stationary states of the system. Exactly as in the case of phonons and tight-binding, one-electron functions, we must form linear combinations which are changed only by a phase factor $e^{i\mathbf{k}\cdot\mathbf{m}}$ when a primitive translation $\mathbf{m}$ is made on the coordinates of the system, i.e.,

$$\Phi_l(\mathbf{K}) = N^{-1/2} \sum_j e^{i\mathbf{K}\cdot\mathbf{R}_j} \Phi_l(\mathbf{R}_j). \quad (3.5)$$

The new first-order eigenvalues corresponding to these states are

$$E_l(\mathbf{K}) = \int \Phi_l(\mathbf{K})^* \mathcal{H}_0 \, \Phi_l(\mathbf{K}) \, d\mathbf{r}_1 \cdots d\mathbf{r}_N$$

$$= E_{ii}^l + \sum_{i'} e^{i\mathbf{K}\cdot(\mathbf{R}_{i'} - \mathbf{R}_i)} E_{ii'}^l. \quad (3.6)$$

Matrix elements connecting states of different $\mathbf{K}$ vanish. Note that the index i in Eq. (3.6) is arbitrary and may be chosen such that $\mathbf{R}_i = 0$ for convenience. The set of eigenvalues (3.6) labeled by N different wave vectors $\mathbf{K}$ constitute an "exciton band," whose width is of the order of magnitude of the largest of the matrix elements $E_{ii'}^l$. E_{ii}^l can be computed on the one-electron-per-atom model, and is found to consist of the ground-state energy in first order, plus the change in energy of the system when a particular atom is excited (approximately equal to an atomic excitation energy). $E_{ii'}^l$ is related to the probability that the excitation energy will jump between the atoms at $\mathbf{R}_i$ and $\mathbf{R}_{i'}$.

Results of somewhat more general usefulness can be obtained on our idealized two-electron-per-atom model. Frenkel's trial states can be viewed as a very special case of Eq. (2.22) in which the localized

---

[42] These matrix elements were called V_{kl} by Frenkel,[1] J_{mn} by Peierls,[3] and E_{JL} by Heller and others.[39–41]

functions $a_{m\mathbf{R}\sigma}$ are by construction true atomic functions, $\phi_{0\mathbf{R}\sigma}$, and in which the excited electron has remained in the cell at $\mathbf{R}_i$ in an excited orbital $\phi_{l\mathbf{R}_i\sigma'}$. Hence, a state in our idealized model which corresponds to the simpler Frenkel trial function (3.2) is

$$\Phi_{0l}^{\sigma\sigma'}(\mathbf{R}_i, \mathbf{R}_i) = \mathscr{A}\phi_{0\mathbf{R}_1\alpha}\phi_{0\mathbf{R}_1\beta}\cdots \underbrace{\phi_{0\mathbf{R}_i\sigma}\phi_{l\mathbf{R}_i\sigma'}}_{\text{wave function of an excited atom at }\mathbf{R}_i}\cdots \phi_{0\mathbf{R}_N\beta}.$$

The Frenkel exciton state is then $\Phi_{0l}(\mathbf{K}, 0)$ [cf. Eqs. (2.28) and (3.5)], and Eqs. (2.24) through (2.29) may now be used, taking $\mathbf{R}_i = \mathbf{R}_j$ and $\mathbf{R}_{i'} = \mathbf{R}_{j'}$ everywhere ($\beta = \beta' = 0$), to compute the first-order Frenkel Hamiltonian matrix. The result, which is taken directly from Eq. (2.35), is

$$H_{ll'}(\mathbf{K}) \equiv \int \Phi_{0l}(\mathbf{K})^* \mathscr{H}_0\, \Phi_{0l'}(\mathbf{K})\, d\tau_1 \cdots d\tau_{2N}$$

$$= \delta_{ll'}E_0^{(1)} + W_{ll'}$$

$$+ \sum_{\mathbf{R}\neq 0} e^{i\mathbf{K}\cdot\mathbf{R}}[2\delta_M(l\mathbf{O}\ 0\mathbf{R}\,|\,g\,|\,0\mathbf{O}\ l'\mathbf{R}) - (l\mathbf{O}\ 0\mathbf{R}\,|\,g\,|\,l'\mathbf{R}\ 0\mathbf{O})] \quad (3.7)$$

where

$$W_{ll'} = W(l'\mathbf{O}, l\mathbf{O}) - W(0\mathbf{O}, 0\mathbf{O})\,\delta_{ll'} + 2\delta_M(l\mathbf{O}\ 0\mathbf{O}\,|\,g\,|\,0\mathbf{O}\ l'\mathbf{O})$$
$$- (l\mathbf{O}\ 0\mathbf{O}\,|\,g\,|\,l'\mathbf{O}\ 0\mathbf{O}) \quad (3.8)$$

which vanishes in the case $l \neq l'$ whenever the crystal point symmetry is such that the atomic states ϕ_l and $\phi_{l'}$ do not mix. This would be the case, for example, if l and l' were to refer to two different p functions in a cubic crystal. Here we have used a boldface **O** to indicate the position of an atom located at the origin, so that $0\mathbf{O}$ stands for a ground-state, one-electron atomic function centered at the origin, and, e.g., $l\mathbf{R}$ stands for an excited state ϕ_l at $\mathbf{R}$. The (diagonal) third term of Eq. (3.7) corresponds to the second term of (3.6). As in Section 2, $\delta_M = 1$ in a singlet state and $\delta_M = 0$ in a triplet state.

The **K**-dependent term in Eq. (3.7) looks at first like a typical tight-binding lattice sum. When $l = l'$ it can, in fact, be interpreted loosely as a constant plus a kinetic energy of the exciton. In any case, considerable care must be taken in performing the lattice sum. Observe that the second term in square brackets,

$$- (l\mathbf{O}\ 0\mathbf{R} \mid g \mid l'\mathbf{R}\ 0\mathbf{O}) \tag{3.9}$$

is an exchange-like interaction which can be regarded as the Coulomb interaction of two overlap charge clouds $\phi_l(\mathbf{r})^* \phi_{l'}(\mathbf{r} - \mathbf{R})$ and $\phi_0(\mathbf{r} - \mathbf{R})^* \phi_0(\mathbf{r})$. As $\mathbf{R}$ is increased, the effective charge of each of these clouds will decrease exponentially, so the term itself will decrease exponentially. However, the first term, which appears only in singlet states,

$$2\ (l\mathbf{O}\ 0\mathbf{R} \mid g \mid 0\mathbf{O}\ l'\mathbf{R}) \tag{3.10}$$

is effectively the Coulomb interaction between the two charge clouds $\phi_l(\mathbf{r})^* \phi_0(\mathbf{r})$ and $\phi_0(\mathbf{r} - \mathbf{R})^* \phi_l(\mathbf{r} - \mathbf{R})$. As R is increased, these clouds cease to overlap with *each other* but Eq. (3.10) does *not* vanish exponentially with R. It remains as the interaction between the multipole moments of these two clouds. This term is usually called an "excitation transfer" interaction because it is effectively a transition matrix element between states in which initially the atom at $\mathbf{O}$ is excited and, finally, the one at $\mathbf{R}$ is excited.

A study of the electronic structure of the exciton in specific systems on the Frenkel model in principle merely involves a computation and diagonalization of the matrix (3.7). In the remainder of this section we review some of the more general aspects of such calculations.

b. *Transverse and Longitudinal Excitons*

Consider the evaluation of the excitation transfer interaction (3.10) in the simple but important case in which the ground state ϕ_0 is s-like and there exist three possible p-like excited states labeled by l. Let $g = e^2/r_{12}$ be expanded in a Taylor series in powers of the components of $\mathbf{r}_1$ and $\mathbf{r}_2 - \mathbf{R}$ (exactly as in the usual treatment of the van der Waals interaction).[39] Because of the orthogonality of ϕ_0 and ϕ_l, only one term of this expansion contributes to Eq. (3.10), i.e.,

$$[(\boldsymbol{\mu}_{l0} \cdot \boldsymbol{\mu}_{0l'})\ R^2 - 3\ (\boldsymbol{\mu}_{l0} \cdot \mathbf{R})\ (\boldsymbol{\mu}_{0l'} \cdot \mathbf{R})]\ R^{-5} \tag{3.11}$$

where

$$\boldsymbol{\mu}_{l0} = 2^{1/2}\ e \int \phi_l(\mathbf{r})^*\ \mathbf{r}\ \phi_0(\mathbf{r})\ d\mathbf{r} \tag{3.12}$$

and

$$\mu_{0l'} = 2^{1/2} e \int \phi_0(\mathbf{r})^* \, \mathbf{r} \, \phi_{l'}(\mathbf{r}) \, d\mathbf{r}. \tag{3.13}$$

The term in square brackets in Eq. (3.7) therefore drops off as R^{-3} at large distances and contains a short-range part which drops off roughly as $e^{-R/d}$, where d characterizes the spatial extent of the charge densities $\phi_l^* \phi_0$ and $\phi_0^* \phi_{l'}$. The short-range part contains the exchange-like term (3.9) and any contribution to Eq. (3.10) not accounted for by its long-range part (3.11).

Let us now further restrict the discussion to cubic crystals, ignore the short-range part of the lattice sum in Eq. (3.7), and examine

$$\sum_{\mathbf{R} \neq 0} e^{i\mathbf{K}\cdot\mathbf{R}} \left[(\mathbf{\mu}_{l0} \cdot \mathbf{\mu}_{0l'}) R^2 - 3 (\mathbf{\mu}_{l0} \cdot \mathbf{R})(\mathbf{\mu}_{0l'} \cdot \mathbf{R}) \right] R^{-5}. \tag{3.14}$$

This term vanishes identically when $\mathbf{K}$ is set equal to zero, since it then represents the interaction between a dipole at the origin and a perfect cubic lattice of dipoles identical to each other (and to the one at the origin if $l = l'$). When $\mathbf{K} \neq 0$, this sum must be computed using methods developed by Born and Bradburn,[43] carried out in detail for cubic crystals by Cohen and Keffer.[44] The result, for $0 < Ka \ll 1$, where a is the lattice constant, is

$$-\frac{4\pi}{3} \rho \mu_{l0} \mu_{0l'} (\delta_{ll'} - 3K_l K_{l'} K^{-2}) + O(K^2 a^2) \tag{3.15}$$

where ρ is the density of lattice points and K_l and $K_{l'}$ are the projections of $\mathbf{K}$ on the respective dipole moments $\mathbf{\mu}_{0l}$ and $\mathbf{\mu}_{0l'}$. As $\mathbf{K} \to 0$, Eq. (3.15) does not approach a unique value. It is piecewise discontinuous at $\mathbf{K} = 0$.[44] We must now choose the best set of p functions with which to work. This choice will in turn determine the orientations of the $\mathbf{\mu}_{l0}$. When a single impurity atom is placed in a static cubic field, and excited to a P level, the three p functions may

---

[43] M. Born and M. Bradburn, *Proc. Cambridge Phil. Soc.* **39**, 104 (1943).

[44] M. H. Cohen and F. Keffer, *Phys. Rev.* **99**, 1128 (1955); Dipole-dipole sums behave much differently in more highly anisotropic crystals; see D. Fox and S. Yatsiv, *Phys. Rev.* **108**, 938 (1957). Further details on dipole-dipole sums in cubic crystals are given by A. A. Demidenko, *Fiz. Tverd. Tela* **3**, 1164 (1961) [see *Soviet Phys.—Solid State (English Transl.)* **3**, 869 (1961)], and by V. N. Piskovoi, *Fiz. Tverd. Tela* **4**, 1393 (1962) [see *Soviet Phys.—Solid State (English Transl.)* **4**, 1025 (1962)].

be oriented in any direction as long as they remain orthogonal. Thus $W_{ll'}$ will be independent of our choice in this problem. Now it is clear that Eq. (3.15) will be extremely simple if ϕ_l and $\phi_{l'}$ are chosen parallel and/or perpendicular to **K**, i.e., if the p functions are made "longitudinal" and "transverse" to **K** (as indicated by the $x'y'z'$ axes in Fig. 6). In fact, except for the terms of order $(Ka)^2$, this

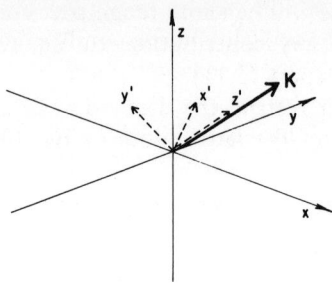

Fig. 6. A coordinate transformation useful in describing transverse and longitudinal states. The xyz axes are asssociated with the crystal symmetry directions. **K** is in an arbitrary direction. z' is parallel to **K**, x' lies in the xy plane, and $x'y'z'$ form an orthogonal system.

choice directly diagonalizes Eq. (3.15). If $l = l'$ and ϕ_l is longitudinal ($\phi_{z'}$), then $K_l = K$ and Eq. (3.15) becomes

$$+ \frac{8\pi}{3} | \mu_{l0} |^2 \rho + O(K^2 a^2) \quad \text{(longitudinal } p\text{-like excitons).} \quad (3.16)$$

If $l = l'$ and ϕ_l is transverse to **K** ($\phi_{x'}$ or $\phi_{y'}$), then by definition $K_l = 0$ and Eq. (3.15) becomes

$$- \frac{4\pi}{3} | \mu_{l0} |^2 \rho + O(K^2 a^2) \quad \text{(tranverse } p\text{-like excitons).} \quad (3.17)$$

Finally, if $l \neq l'$, then (necessarily) either K_l or $K_{l'}$ or both will be zero and Eq. (3.15) is merely of order $(Ka)^2$. *Longitudinal and transverse exciton states near $K = 0$ built from p functions will thus be split by an amount* $4\pi\rho | \mu_{l0} |^2$, which is proportional to f_{0l}, the atomic $S \to P$ oscillator strength. This contrasts sharply with the short-range contribution to the lattice sum, which always has the form

$$\delta_{ll'} V_l + O(K^2 a^2) \quad (2.18)$$

reminiscent of the situation in the tight binding theory of electronic p

bands.[45] The short-range contribution to the exciton energy therefore produces no longitudinal-transverse splitting.

The longitudinal-transverse effect in excitons was discovered by Heller and Marcus,[39] and it has been discussed frequently in recent years.[46–49] Although it is most easily demonstrated in detail on the Frenkel model, it is a general phenomenon, expected to occur in any set of exciton states which can be produced by an allowed (electric dipole) transition. The splitting corresponds exactly to the Lyddane-Sachs-Teller splitting of the longitudinal and transverse optical branches near the center of the zone in the phonon spectrum.[50] Hopfield and Thomas[49] have developed a phenomenological theory of longitudinal and transverse excitons in cubic and uniaxial crystals using Born and Huang's virtual oscillator concept. As we will see in detail in Section 9, optical absorption generally produces transverse exciton states because the exciton's momentum **K** must match that of the absorbed photon, whose electric field is transverse. The region near $\mathbf{K} = 0$ is of special interest because the size of the propagation vector of light is of the order of 10^{-3} reciprocal lattice vectors.

Transverse and longitudinal exciton states have been shown above to be uncoupled in a cubic crystal when $Ka \ll 1$, regardless of the direction of **K**. This is an "accidental" uncoupling, for as far as rigorous symmetry arguments are concerned, decoupling is expected only when the transverse and longitudinal states transform according to different irreducible representations of the group of **K**.[51,52] In a cubic crystal this is true for an arbitrarily large magnitude of **K** only if **K** lies in one of the directions of high symmetry. For example, the (100) longitudinal exciton transforms according to $\varDelta_1$, while the transverse exciton is $\varDelta_5$-like. We emphasize that longitudinal-transverse mixing cannot be completely ignored in cubic crystals, since excitons of large K may sometimes be involved in "indirect" transitions (Section 10). In noncubic crystals this mixing can even

---

[45] F. Seitz, "Modern Theory of Solids," p. 306. McGraw-Hill, New York, 1940.
[46] U. Fano, *Phys. Rev.* **103**, 1202 (1956); **118**, 451 (1960).
[47] V. L. Bonch-Bruevich, *Fiz. Metal. i Metalloved.* **4**, 546 (1957).
[48] S. I. Pekar, *Zh. Eksperim. i Teor. Fiz.* **35**, 522 (1958); see *Soviet Phys. J.E.T.P.* **8**, 360 (1959).
[49] J. J. Hopfield and D. G. Thomas, *Phys. Chem. Solids* **12**, 276 (1960).
[50] See, e.g., M. Born and K. Huang, "Dynamical Theory of Crystal Lattices," p. 87. Oxford Univ. Press, London and New York, 1954.
[51] L. P. Bouckaert, R. Smoluchowski, and E. P. Wigner, *Phys. Rev.* **50**, 58 (1936).
[52] G. F. Koster, *Solid State Phys.* **5**, 173 (1957).

occur at small values of K as soon as the propagation vector moves away from a direction of high symmetry. Hopfield and Thomas have used this fact to observe longitudinal excitons by virtue of a small transverse admixture in experiments on the hexagonal crystals ZnO[49] and CdS.[53]

Summarizing the results of this subsection, we write the energy of a transverse singlet Frenkel exciton state relative to the first-order, ground-state energy as

$$W_{11} - \frac{4\pi}{3} \rho \, | \, \mu_{01} \, |^2 + V_1 + O(K^2 a^2) \qquad (3.19)$$

and that of the longitudinal singlet as

$$W_{11} + \frac{8\pi}{3} \rho \, | \, \mu_{01} \, |^2 + V_1 + O(K^2 a^2) \, . \qquad (3.20)$$

The term involving $| \, \mu_{01} \, |^2$ is missing in triplet states. Here W_{11} and V_1 are the values of $W_{ll'}$ and V_l [Eqs. (3.8) and (3.18), respectively] computed with any one of the three excited p functions. The terms of order $K^2 a^2$ are different for the two cases (3.19) and (3.20) and have not been given an explicit formulation because of their complexity and dependence on the specific cubic structure involved. The coefficients of $K^2 a^2$ can be written in terms of an exciton effective mass, M^*, for which Heller and Marcus gave the approximate expression

$$\frac{m}{M^*} = - \delta \left[\frac{16\pi \, (\frac{3}{4} \pi)^{2/3}}{45} \right] \frac{e^2 \rho f_{01}}{\epsilon_1 - \epsilon_0} \qquad (3.21)$$

where $\delta = -\frac{1}{2}$ for transverse excitons and $+1$ for longitudinal excitons, f_{01} is the oscillator strength for the atomic $s \to p$ transition, m is the electron mass, and $\epsilon_1 - \epsilon_0$ is the atomic excitation energy. This expression, which is often quoted, ignores the short-range contribution to the lattice sum. This will be discussed in the following. The full **K** dependence of the exciton energies can be obtained only by performing the various lattice sums, using tight-binding methods for the short-range term and the methods of Born and Bradburn[43]

---

[53] J. J. Hopfield and D. G. Thomas, *Phys. Rev.* **122**, 35 (1961); The longitudinal-transverse splitting has also been observed in CdS by E. F. Gross and B. S. Razbirin, *Fiz. Tverd. Tela* **4**, 207 (1962) [*English transl.: Sov. Phys. Solid State* **4**, 146 (1962)].

and Cohen and Keffer[44] for the long-range term. In Fig. 7 we give examples of the long-range lattice sum for the face-centered cubic lattice.

Since it arises from the long-range interaction, the longitudinal-transverse splitting is absent in purely triplet levels [cf. Eq. (3.7)].

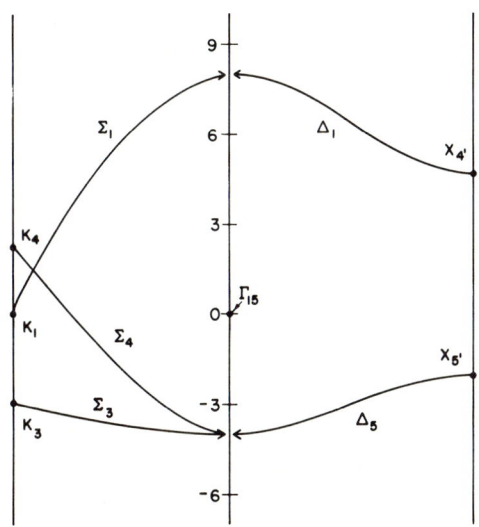

FIG. 7. The dipole sum [Eq. (3.14)], evaluated for an infinite perfect face-centered cubic crystal [after M. H. Cohen and F. Keffer, *Phys. Rev.* **99**, 1128 (1955)]. Representation notation is that of L. P. Bouckaert, R. Smoluchowski, and E. P. Wigner [*Phys. Rev.* **50**, 58 (1936)].

However, the splitting will appear in triplet levels whenever any appreciable singlet admixture due to spin-orbit interaction occurs. Some care is necessary in interpreting data on thin samples, since the splitting is less well-defined near $\mathbf{K} = 0$ in finite crystals.[44]

c. *Davydov Splitting*[13]

The Frenkel model has been used extensively in the theory of excited states of molecular crystals. Since these crystals (e.g., solid benzene and anthracene) rarely have simple Bravais crystal structures, it is necessary to generalize the simple Frenkel theory slightly to obtain a useful formalism. We restrict our discussion to the most elementary aspects of this subject, since reviews of work on excited states of molecular crystals have recently appeared in *Solid State Physics*.[14,15]

Consider a very simple crystal containing two molecules per unit cell, such that each molecule sees the same environment. An example of this in two dimensions is sketched in Fig. 8. Molecules A and B are

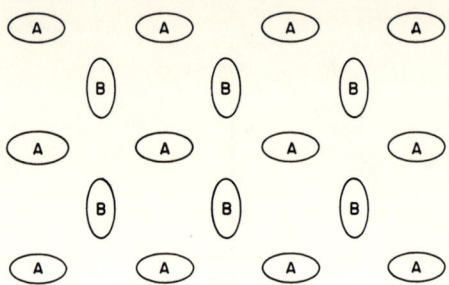

FIG. 8. Idealized two-dimensional molecular crystal. Molecules A and B are identical but oriented differently.

identical but oriented differently. Molecule A is located at the origin in the unit cell and molecule B is displaced from it by a basis vector **b**.

Zero-order ground and excited states of this system are easily constructed by analogy with those of the idealized model. We fix our attention on ground-state and excited-state molecular orbitals ϕ_0 and ϕ_l, and then make up a ground-state Φ_0 as an antisymmetrized product of all ground-state orbitals on molecules of both types. Excited states $\Phi_{0,Al}^{\sigma\sigma'}(\mathbf{R}, \mathbf{R})$ and $\Phi_{0,Bl}^{\sigma\sigma'}(\mathbf{R} + \mathbf{b}, \mathbf{R} + \mathbf{b})$ are constructed as antisymmetrized products of the same sets of orbitals, but with *either* an "A" molecule at $\mathbf{R}$ *or* a "B" molecule at $\mathbf{R} + \mathbf{b}$, respectively, in the excited state $\phi_{0\sigma}\phi_{l\sigma'}$. Only those localized excitations constructed on a given type of molecule (A or B) are translationally equivalent, so two separate Frenkel excitons must be considered. They are

$$\Phi_{Al}(\mathbf{K}) = N^{-1/2} \sum_{\mathbf{R}} e^{i\mathbf{K}\cdot\mathbf{R}} \Phi_{Al}(\mathbf{R}) \quad (3.22)$$

and

$$\Phi_{Bl}(\mathbf{K}) = N^{-1/2} \sum_{\mathbf{R}} e^{i\mathbf{K}\cdot\mathbf{R}} \Phi_{Bl}(\mathbf{R} + \mathbf{b}). \quad (3.23)$$

For brevity we have written $\Phi_{Al}(\mathbf{R})$ and $\Phi_{Al}(\mathbf{K})$ for $\Phi_{0,Al}^{\sigma\sigma'}(\mathbf{R}, \mathbf{R})$ and $\Phi_{0,Al}^{\sigma\sigma'}(\mathbf{K}, 0)$, respectively, suppressing irrelevant indices. The expectation value of $\mathscr{H}_0$, which now includes all interactions among the $4N$ valence electrons and $2N$ cores of the molecules, is given by Eq. (3.7), in which W_{ll} is now the energy required to excite a single molecule locally in the presence of the rest of the entire crystal (not just those molecules on its translationally equivalent sublattice). Thus

$$\int \Phi_{Al}(\mathbf{K})^* \mathscr{H}_0 \, \Phi_{Al}(\mathbf{K}) \, d\tau_1 \cdots d\tau_{4N} = E_0^{(1)} + W_{ll}$$
$$+ \sum_R e^{i\mathbf{K}\cdot\mathbf{R}} \left[2\delta_M (Al\mathrm{O}\ A0\mathrm{R} \mid g \mid A0\mathrm{O}\ Al\mathrm{R}) - (Al\mathrm{O}\ A0\mathrm{R} \mid g \mid Al\mathrm{R}\ A0\mathrm{O}) \right] \quad (3.24)$$

and

$$\int \Phi_{Bl}(\mathbf{K})^* \mathscr{H}_0 \, \Phi_{Bl}(\mathbf{K}) \, d\tau_1 \cdots d\tau_{4N} = E_0^{(1)} + W_{ll}$$
$$+ \sum_R e^{i\mathbf{K}\cdot\mathbf{R}} \left[2\delta_M (Bl\mathrm{O}\ B0\mathrm{R} \mid g \mid B0\mathrm{O}\ Bl\mathrm{R}) - (Bl\mathrm{O}\ B0\mathrm{R} \mid g \mid Bl\mathrm{R}\ B0\mathrm{O}) \right]. \quad (3.25)$$

Coordinates in the matrix elements in Eq. (3.25) have been shifted by an amount $-\mathbf{b}$ to simplify the notation. Physically, Eqs. (3.24) and (3.25) express the energies of states in which Frenkel excitons travel independently on the A and B sublattices, paying no attention to each other. It is clear, however, that a localized excitation might want to jump from an A molecule to a B molecule. The molecules are identical, and are in identical surroundings, so there exists a condition of resonance. $\mathscr{H}_0$ contains interactions among all electrons, and therefore couples the two trial states (3.22) and (3.23). The coupling matrix element is

$$\int \Phi_{Al}(\mathbf{K})^* \mathscr{H}_0 \, \Phi_{Bl}(\mathbf{K}) \, d\tau_1 \cdots d\tau_{4N}$$
$$= \sum_R e^{i\mathbf{K}\cdot\mathbf{R}} \left[2\delta_M (Al\mathrm{O}\ B0\mathrm{R}+\mathbf{b} \mid g \mid A0\mathrm{O}\ Bl\mathrm{R}+\mathbf{b}) \right.$$
$$\left. - (Al\mathrm{O}\ B0\mathrm{R}+\mathbf{b} \mid g \mid Bl\mathrm{R}+\mathbf{b}\ A0\mathrm{O}) \right]. \quad (3.26)$$

Let us refer to Eqs. (3.24) through (3.26) as $E_{AA}^l(\mathbf{K})$, $E_{BB}^l(\mathbf{K})$, and $E_{AB}^l(\mathbf{K})$, respectively. Then the true first-order eigenvalues of $\mathscr{H}_0$ corresponding to the lth excited trial state are

$$\tfrac{1}{2}[E_{AA}^l(\mathbf{K}) + E_{BB}^l(\mathbf{K})] \pm \{\tfrac{1}{4}[E_{AA}^l(\mathbf{K}) - E_{BB}^l(\mathbf{K})]^2 + |E_{AB}^l(K)|^2\}^{1/2}. \quad (3.27)$$

Associated with these eigenvalues are two orthogonal linear combinations of $\Phi_{Al}(\mathbf{K})$ and $\Phi_{Bl}(\mathbf{K})$ which are quite simple in the event that $E_{AA}^l = E_{BB}^l$, viz.,

$$2^{-1/2}\left[\Phi_{Al}(\mathbf{K}) \pm \Phi_{Bl}(\mathbf{K})\right] \tag{3.28}$$

in which case Eq. (3.27) reduces to

$$E_{AA}^l(\mathbf{K}) \pm E_{AB}^l(\mathbf{K}). \tag{3.29}$$

This result may be compared with Eq. (1.13) of reference 14.

The polarization of the molecular transition $\phi_0 \to \phi_l$ relative to any arbitrary axis will depend on the orientation of the molecule under consideration. In general, then, $E_{AA}^l(\mathbf{K})$ will not necessarily be equal to $E_{BB}^l(\mathbf{K})$, since the lattice sums in Eqs. (3.24) and (3.25) may depend on the relative orientations of molecules A and B with respect to $\mathbf{K}$. $E_{AA}^l(\mathbf{K})$ is not necessarily equal to $E_{BB}^l(\mathbf{K})$ even at $\mathbf{K} = 0$ if the "site symmetries" of A and B are not the same, i.e., if A and B molecules do not see precisely the same environment. Regardless of these considerations, the splitting displayed in Eq. (3.27) persists because of E_{AB}^l. This splitting, often attributed to E_{AB}^l alone [as in Eq. (3.29)] because E_{AB}^l makes the largest contribution, is known as "Davydov" or "factor group" splitting. This effect is peculiar to crystalline aggregates of molecules since its detection depends on the existence of a large number of pairs (or groups) of molecules oriented in some specific way with respect to one another. Optical transitions to both of the states (3.28) are generally possible; when there are more than two (say n) translationally inequivalent molecules, there will be at most n levels in the Davydov multiplet. The effect has been observed in the absorption spectra of a number of crystals (e.g., benzene,[54] naphthalene,[55] anthracene,[56] and CO[57]). Experiments on polarizations and intensities of the observed lines are reviewed thoroughly in the articles by McClure and Wolf.

The assumption of nondegeneracy in the excited molecular state, which has been made throughout this discussion, is not a severe restriction on the theory. The relatively low symmetry of molecules

[54] V. L. Broude, V. S. Medvedev, and A. F. Prikhot'ko, *Zh. Eksperim. i Teor. Fiz.* **21**, 673 (1951).

[55] A. F. Prikhot'ko, *Zh. Eksperim. i Teor. Fiz.* **19**, 383 (1949); D. S. McClure and O. Schnepp, *J. Chem. Phys.* **23**, 1575 (1955).

[56] D. P. Craig and P. C. Hobbins, *J. Chem. Soc.* pp. 539, 2302, and 2309 (1955).

[57] K. Dressler, *J. Quant. Spectry. Radiative Transfer* **2**, 683 (1962).

provides us with many more interesting nondegenerate excited states than are available in spherically symmetric atoms. Winston[58] has given a general treatment of the case of degenerate excited molecular states in crystals.

Recent studies of molecular crystal excitons have been increasingly concerned with the possibility that triplet excitons, presumably having lifetimes sufficiently long, lead to observable magnetic effects because of their nonzero spin.[59-61]

d. Frenkel Excitons in Solid Argon

The solid rare gases have long been considered[3,38,41] ideally suited for application of the Frenkel model because of their relatively tight binding in their ground states, their cubic symmetry, and their relatively large lattice constants ($\sim 10a_0$). In 1958 the author[23,62] computed the excitation energies and band structure of low-lying Frenkel excitons in solid argon, and will present here a qualitative account of this study. Experimental work on the absorption spectra of solid rare gases has appeared only in the past few years.[22,63,64]

Atomic argon has a ground-state configuration $1s^2 \cdots 3p^6$ and its lowest lying excited states are $1s^2 \cdots 3p^5 4s(^1P, \,^3P)$. Zero-order ground- and excited-state functions in the solid of the general nature of Eq. (2.5) were used, including all core and valence electrons explicitly instead of just two per atom. If purely atomic functions were inserted in these products, the resulting functions would not be normalized because of overlapping between atomic functions on different sites. Typical overlaps in solid argon for nearest neighbors are -0.027 ($3p\sigma - 3p\sigma$) and -0.081 ($3p\sigma - 4s\sigma$). Löwdin's technique of symmetric orthogonalization[31,32] was therefore used to modify the atomic

---

[58] H. Winston, *J. Chem. Phys.* **19**, 156 (1951).
[59] H. Sternlicht and H. M. McConnell, *J. Chem. Phys.* **35**, 1793 (1961).
[60] H. M. McConnell and R. Lynden-Bell, *J. Chem. Phys.* **36**, 2393 (1962).
[61] H. M. McConnell, H. O. Griffith, and D. Pooley, *J. Chem. Phys.* **36**, 2518 (1962).
[62] R. S. Knox, Thesis, University of Rochester, 1958 (unpublished).
[63] J. R. Nelson and P. L. Hartman, *Bull. Am. Phys. Soc.* [2] **4**, 371 (1959), appear to have done the first experiments on solid argon spectra (at Cornell University in 1957). However, at least one of their strong absorption bands (at 9 ev) has not been confirmed by other workers.[22,64]
[64] G. Baldini, *Phys. Rev.* **128**, 1562 (1962); in private communication, Baldini informs the author that Xe impurities in lighter rare gas solids appear to display Wannier series with reduced masses characteristic of the host conduction electrons.

functions before constructing states (2.5). The term W_{ll} in Eq. (3.7), which we have seen to be a perturbed atomic excitation energy, was computed in an approximation in which all terms of higher order than second in overlap quantities were neglected.[31] The results were that the atomic $^1P_1$ and $^3P_1$ states at 11.8 and 11.6 ev, respectively, should contribute a similar pair of states centered at 9 ($\pm$ 1) ev in the solid. This large probable computational error was a result of approximations made in evaluating certain two-electron and three-center integrals, and in trusting the atomic Hartree-Fock wave functions as extrapolated to large distances. While one might regard the atomic levels as being simply "perturbed" by 2.6 ev, this is illusory. There occur enormous contributions to W_{ll} which very nearly cancel each other. The positive and negative corrections to the atomic excitation energy add up separately to $+$ 8.3 and $-$ 10.9 ev. Of the positive correction, nearly one-half is contributed by three-center integrals. Indeed, the Frenkel model must strain itself to accommodate this "ideally suited" crystal. The 4s electron is just not very tightly bound.

Baldini's data[64] reveals strong absorption by solid argon at about 12.2 ev. Moreover, in all the solid rare gases absorption lines occur very close to those of the free atoms. A probable cause of the discrepancy between the above theory and the experiments is the poor convergence of the Löwdin orthogonalization in the excited states; the computation of W_{ll} might not even converge at any order in S (overlap quantities), whereas it was cut off at order S^2. Gold, encountering similar difficulties in an impurity center calculation,[65] has proposed[66] that the extraordinarily large overlap of the argon 4s function with its neighbors might best be handled by exact Schmidt orthogonalization. His Schmidt calculations on an argon impurity in solid neon[66] indicate that this procedure may raise the predicted excitation energy somewhat, and a recalculation for pure argon along these lines will probably be very useful. The possibility that the Wannier model is at least as good as the Frenkel model in the rare gases has been raised recently by Baldini[64] and the author.[67]

A second part of the solid argon calculation was a complete treatment of the **K**-dependent terms in Eqs. (3.8) and (3.11). Although

---

[65] A. Gold, *Phys. Chem. Solids* **18**, 218 (1961).

[66] A. Gold, *Phys. Rev.* **124**, 1740 (1961).

[67] R. S. Knox, *Radiation Res., Suppl.* in press *(Proc. Colo. Springs Symp. on Excitons, May 1962)*.

there are no experimental data which bear on this band structure, one feature is worth noting because of a conjecture by Heller and Marcus[39] that the long-range (excitation-transfer) interaction will generally outweigh the short-range interactions in contributing to the reciprocal effective exciton mass. The opposite is true in solid argon, and may well be true for the Frenkel picture of the alkali halides to which their remarks applied. The main reason for this reversal is the enormous overlap of excited $4s$ electrons situated on nearest neighbors ($S_{4s,4s} = 0.55$). Overlap charge densities of this

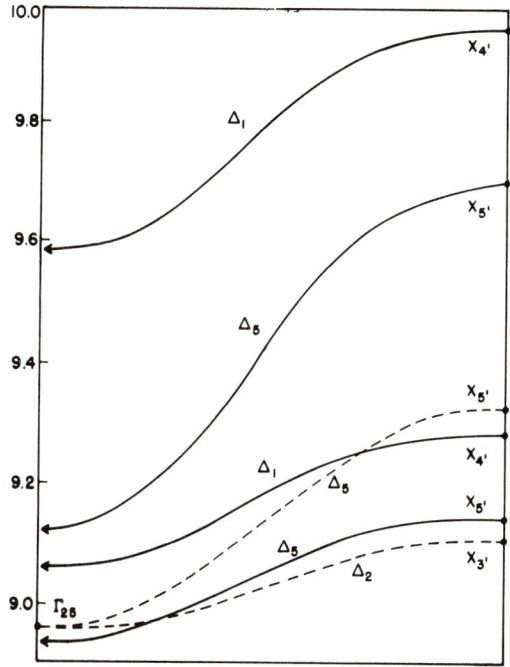

FIG. 9. Exciton bands in solid argon in the (100) direction. Notation is that of L. P. Bouckaert, R. Smoluchowski, and E. P. Wigner [*Phys. Rev.* **50**, 58 (1936)], and the ordinate is in ev measured relative to the first-order, ground-state energy of the crystal. Spin-orbit interaction has been included, and the two bands drawn with a solid line marked Δ_5 are transverse excitons (the lower one mostly singlet, the upper mostly triplet). The Δ_1 bands are longitudinal excitons (the upper mostly singlet, the lower mostly triplet). The dashed lines show bands derived from the atomic $^3P_2$ excited state which are optically uninteresting. Of these the Δ_5 state nevertheless affects the width of the other (transverse) bands, which are derived from the $^3P_1$ and $^1P_1$ states. [After R. S. Knox, *Phys. Chem. Solids* **9**, 265 (1959).]

type do not enter into the computation of W_{ll} but do enter into exchange-type interactions like Eq. (3.13). Another factor contributing to the smaller importance of the long-range interaction here is the fact that the argon $3p \rightarrow 4s$ oscillator strength is only about 0.2, rather than unity, as assumed for alkali halides by Heller and Marcus. The effective mass of the Frenkel exciton in argon increases by a factor of about 5 (from 5 to 25 electron masses) when all short-range terms are dropped from the calculation.

The Frenkel exciton band structure in argon in the (100) direction in $\mathbf{K}$ space is sketched in Fig. 9. Note that while the long-range interaction does produce longitudinal-transverse splitting (states Δ_1 and Δ_5 do not converge to Γ_{15} at $\mathbf{K} = 0$), the short-range interaction is strong enough to change the sign of the longitudinal exciton's effective mass (cf. Fig. 8).

e. Range of Validity and Usefulness

A reasonably quick solution of the Frenkel exciton problem has been obtained because of the assumption that the excited electron primarily shares a unit cell with the hole. There is no problem of diagonalizing the matrix (2.29) with respect to β, because the set of trial wave functions does not include any for which $β \neq 0$. This severely limits the usefulness of the model. Formally, it is possible to solve any exciton problem with Frenkel states. A very large radius exciton would then be an enormously complicated mixture of highly interacting, high-energy atomic states. Large-radius excitons find a much more natural description in terms of the Wannier model, as shown in Section 4.

Physically, a single Frenkel exciton will be a good eigenstate of $\mathcal{H}_0$ if it can be correctly argued that it lies well below all higher excitations and does not interact strongly with them. This is probably undisputably true only of molecular excitons, since excitation of a molecule often involves merely "rearrangement" of atomic orbitals with no increase in the over-all overlap with neighboring molecules. Obviously, in spite of solid argon's tight binding in the ground state, there is considerable opportunity for the spatially diffuse $4s$ electron to spend time outside the unit cell from which it came. The question of the necessity of including higher states at $β = 0$ or other states with $β \neq 0$ is thus opened. On a strict Heitler-London basis, the latter would mean allowing the $4s$ electron to go into $Ar^-(3p^64s)$ "electron transfer" states on neighboring argon atoms. This is impractical to implement, again because of large overlap integrals, and because it would require some annoying generalizations of the formalism of Section 2. Alternatives will be considered in Sections 4 and 5.

4. Wannier's Case: Weak Binding[5,20,21]

a. Eigenvalues and Eigenfunctions

A rough picture of the large-radius exciton may be obtained by ignoring all particles in the system except the excited electron and hole. The effect of the remaining valence electrons and cores is to provide a periodic potential which gives these particles (say) isotropic effective masses m_e^* and m_h^*, respectively. If it is assumed, furthermore, that the electron and hole interact according to the Coulomb law as modified by the medium, $-e^2/\epsilon r_{eh}$, the Schrödinger equation of the two-particle system is

$$\left(-\frac{\hbar^2 \nabla_e^2}{2m_e^*} - \frac{\hbar^2 \nabla_h^2}{2m_h^*} - \frac{e^2}{\epsilon r_{eh}}\right)\Psi = E\Psi. \quad (4.1)$$

Here ϵ is a dielectric constant whose choice will be discussed later. It has been assumed that the conduction and valence bands are nondegenerate and of standard form. It is instructive to separate this equation by assuming that Ψ is of the form

$$\Psi(\mathbf{R}, \mathbf{r}) = g(\mathbf{R})f(\mathbf{r}) \quad (4.2)$$

where

$$\mathbf{R} = \tfrac{1}{2}(\mathbf{r}_e + \mathbf{r}_h) \quad (4.3)$$

and

$$\mathbf{r} = \mathbf{r}_e - \mathbf{r}_h \quad (4.4)$$

are the average electron-hole coordinate and the electron-hole separation, respectively. The gradient operators corresponding to $\mathbf{R}$ and $\mathbf{r}$ are $\nabla_\mathbf{R} = \nabla_e + \nabla_h$ and $\nabla = \tfrac{1}{2}(\nabla_e - \nabla_h)$, respectively, so in the new coordinate system the two-particle Schrödinger equation (4.1) becomes

$$\left[\tfrac{1}{8}\left(\frac{1}{m_e^*} + \frac{1}{m_h^*}\right)p_\mathbf{R}^2 - \tfrac{1}{2}\left(\frac{1}{m_h^*} - \frac{1}{m_e^*}\right)\mathbf{p}_\mathbf{R}\cdot\mathbf{p}\right.$$
$$\left. + \tfrac{1}{2}\left(\frac{1}{m_e^*} + \frac{1}{m_h^*}\right)p^2 - \frac{e^2}{\epsilon r}\right]\Psi = E\Psi \quad (4.5)$$

where $\mathbf{p_R} = -i\hbar\nabla_\mathbf{R}$ and $\mathbf{p} = -i\hbar\nabla$. Now $g(\mathbf{R})$ can be taken to be $e^{i\mathbf{K}\cdot\mathbf{R}}$ since $p_\mathbf{R}$ commutes with the two-particle Hamiltonian. Making this substitution in Eq. (4.2) and separating (4.5), we find that f must satisfy

$$\left[\frac{p^2}{2\mu} - \frac{e^2}{\epsilon r} - \frac{\hbar}{2}\left(\frac{1}{m_h^*} - \frac{1}{m_e^*}\right)\mathbf{K}\cdot\mathbf{p}\right]f(\mathbf{r}) = \left[E - \frac{1}{8\mu}\hbar^2 K^2\right]f(\mathbf{r}) \qquad (4.6)$$

where μ, defined by

$$\frac{1}{\mu} = \frac{1}{m_e^*} + \frac{1}{m_h^*} \qquad (4.7)$$

is the reduced mass of the pair. The term in $\mathbf{K}\cdot\mathbf{p}$ is analogous to the one encountered in the theory of electronic band structure, and can be eliminated in the present calculation by the use of the transformation

$$f(\mathbf{r}) = e^{i\alpha\mathbf{K}\cdot\mathbf{r}} F(\mathbf{r}) \qquad (4.8)$$

which, substituted in Eq. (4.6), is found to eliminate the $\mathbf{K}\cdot\mathbf{p}$ term provided one takes

$$\alpha = \tfrac{1}{2}(m_e^* - m_h^*)/(m_e^* + m_h^*). \qquad (4.9)$$

Then F must satisfy the simpler equation

$$\left(\frac{p^2}{2\mu} - \frac{e^2}{\epsilon r}\right)F = \left[E - \frac{\hbar^2 K^2}{2(m_e^* + m_h^*)}\right]F. \qquad (4.10)$$

The transformation Eq. (4.8) is equivalent to having orginally chosen a center-of-mass coordinate system in which $\mathbf{r} = \mathbf{r}_e - \mathbf{r}_h$ but in which now

$$\mathbf{R} \to \boldsymbol{\rho} = (m_e^*\mathbf{r}_e + m_h^*\mathbf{r}_h)/(m_e^* + m_h^*). \qquad (4.11)$$

The solutions of Eq. (4.10) are those of a hydrogen atom with an effective Z of $1/\epsilon$, from which we can conclude that for each value of $\mathbf{K}$ there exists a set of bound states at energies

$$E_n(\mathbf{K}) = -\frac{\mu e^4}{2\hbar^2\epsilon^2 n^2} + \frac{\hbar^2 K^2}{2(m_e^* + m_h^*)} \qquad (4.12)$$

4. WANNIER'S CASE: WEAK BINDING

with n^2 states associated with each level labeled by n. The total wave function of the system is

$$\Psi_{nlm,\mathbf{K}} = e^{i\mathbf{K}\cdot(\mathbf{R}+\alpha\mathbf{r})} F_{nlm}(\mathbf{r}) \tag{4.13}$$

where the F_{nlm} are appropriately modified hydrogenic functions ($Z \to 1/\epsilon$) and in which $\mathbf{R} + \alpha\mathbf{r}$ is just the coordinate of the center of mass ρ of the pair. It should be noted further that for every hydrogenic state there is a large number of states labeled by different $\mathbf{K}$ vectors; when a periodic potential is present, this $\mathbf{K}$ will essentially range over the first BZ and an exciton "band" results, exactly as in the Frenkel case. It should be emphasized that these bands represent states of the entire crystal and cannot be placed on a one-electron band structure diagram without danger of confusion.

The zero of energy in Eq. (4.12) has been chosen to correspond to a state in which the electron and hole have been given sufficient energy to free themselves from one another. The exciton "rydberg" is $G = \mu e^4/2\hbar^2\epsilon^2$, the energy required to separate an electron and hole in the state $n = 1$. This state is often called the "ground state of the exciton," but we prefer to avoid this usage. The state $n = 1$ is the lowest *excited* state of the *crystal* in this model.

The two-particle model gives a good account of the main features of many spectra, but it is difficult to generalize. We therefore return to the general formalism of Section 2 and attempt to obtain the preceding results from it. Recall that the matrix of the total Hamiltonian in the exciton representation Eq. (2.29) was left nondiagonal in β, a vector which we introduced as an admittedly poor quantum number and which characterized the spacing between the electron and hole in the exciton state. The Frenkel exciton was obtained by ignoring all matrix elements except those with β = 0. To study the Wannier exciton, we separate $(mn\mathbf{K}\beta | \mathcal{H}_0 | mn\mathbf{K}\beta')$ into two parts, as follows [cf. Eqs. (2.29) and (2.30)]:

$$(mn\mathbf{K}\beta | \mathcal{H}_0 | mn\mathbf{K}\beta') = \mathcal{H}_{\beta\beta'}(\mathbf{K}) + \mathcal{I}_{\beta\beta'}(\mathbf{K}) \tag{4.14}$$

where

$$\mathcal{H}_{\beta\beta'}(\mathbf{K}) = \delta_{\beta\beta'} E_0^{(1)} + e^{i\mathbf{K}\cdot(\beta-\beta')/2} N^{-1/2} \sum_{\mathbf{k}} e^{i\mathbf{k}\cdot(\beta-\beta')}$$
$$\times [W_n(\mathbf{k} + \tfrac{1}{2}\mathbf{K}) - W_m(\mathbf{k} - \tfrac{1}{2}\mathbf{K})] - \delta_{\beta\beta'} e^2/\beta \tag{4.15}$$

and

$$\mathscr{I}_{\beta\beta'}(\mathbf{K}) = -[(n\beta m\mathbf{O}|g|n\beta' m\mathbf{O}) - \delta_{\beta\beta'}e^2/\beta] + 2\delta_M(n\beta m\mathbf{O}|g|m\mathbf{O}n\beta')$$
$$+ \sum_{\mathbf{R}\neq 0} e^{i\mathbf{K}\cdot\mathbf{R}} [2\delta_M(n\beta m\mathbf{R}|g|m\mathbf{O}n\mathbf{R}+\beta')$$
$$- (n\beta m\mathbf{R}|g|n\mathbf{R}+\beta' m\mathbf{O})]. \qquad (4.16)$$

This separation is motivated in part by the observation that those terms collected into $\mathscr{H}_{\beta\beta'}$ are the only ones which cannot be easily shown to be negligible at large values of β. It is proposed to insert $\mathscr{H}_{\beta\beta'}(\mathbf{K})$ into the secular equations (2.32), obtain a new set of wave functions (2.31), and then treat $\mathscr{I}_{\beta\beta'}(\mathbf{K})$ as a perturbation. As we will see shortly, the first phase of this program is Wannier's calculation, leading to the hydrogen-like spectrum of eigenvalues. The second phase, involving $\mathscr{I}_{\beta\beta'}(\mathbf{K})$, was not considered explicitly by Wannier, since for the most part it involves short-range interactions of no consequence to the over-all hydrogenic model.

Inserting $\mathscr{H}_{\beta\beta'}(\mathbf{K})$ into Eq. (2.32), we have

$$E_0^{(1)}U'_{\nu\mathbf{K}}(\boldsymbol{\beta}) + \sum_{\boldsymbol{\beta}'} N^{-1} \sum_{\mathbf{k}} e^{i\mathbf{k}\cdot(\boldsymbol{\beta}-\boldsymbol{\beta}')} [W_n(\mathbf{k}+\tfrac{1}{2}\mathbf{K})$$
$$- W_m(\mathbf{k}-\tfrac{1}{2}\mathbf{K})] U'_{\nu\mathbf{K}}(\boldsymbol{\beta}') - (e^2/\beta) U'_{\nu\mathbf{K}}(\boldsymbol{\beta}) = EU'_{\nu\mathbf{K}}(\boldsymbol{\beta}) \qquad (4.17)$$

where

$$U'_{\nu\mathbf{K}}(\boldsymbol{\beta}) = e^{-i\mathbf{K}\cdot\boldsymbol{\beta}/2} U_{\nu\mathbf{K}}(\boldsymbol{\beta}). \qquad (4.18)$$

Equation (4.17) forms a set of coupled difference equations for the U' which admit of no simple solution. However, Wannier showed that quantities of the general form

$$N^{-1} \sum_{\boldsymbol{\beta}'} \sum_{\mathbf{k}} e^{i\mathbf{k}\cdot(\boldsymbol{\beta}-\boldsymbol{\beta}')} f(\mathbf{k}) g(\boldsymbol{\beta}') \qquad (4.19)$$

could be transformed into differential form as follows. Let $g(\boldsymbol{\beta}')$ have a Fourier transform $G(\varkappa)$, i.e., let

$$g(\boldsymbol{\beta}') = N^{-1/2} \sum_{\varkappa} e^{-i\varkappa\cdot\boldsymbol{\beta}'} G(\varkappa) \qquad (4.20)$$

whereupon Eq. (4.19) becomes

$$N^{-1/2} \sum_{\varkappa} e^{-i\varkappa\cdot\boldsymbol{\beta}} f(\varkappa) G(\varkappa) \qquad (4.21)$$

since $\sum_{\beta'} e^{i(\mathbf{k}+\mathbf{\varkappa})\cdot\mathbf{\beta}'} = N\delta(\mathbf{k}+\mathbf{\varkappa})$. However, for functions $f(\mathbf{\varkappa})$ of practical interest (e.g., those expandable in powers of the components of $\mathbf{\varkappa}$), expression (4.21) is equivalent to

$$f(-i\nabla)g(\mathbf{\beta}) \tag{4.22}$$

(where ∇ operates on the coordinate $\mathbf{\beta}$), as may be verified by a brief calculation. It follows from the equivalence of Eqs. (4.19) and (4.22) that (4.17) is equivalent to

$$[W_n(-i\nabla + \tfrac{1}{2}\mathbf{K}) - W_m(-i\nabla - \tfrac{1}{2}\mathbf{K}) - e^2/\beta]\, U' = (E - E_0^{(1)})\, U'. \tag{4.23}$$

In order to simplify our discussion for the moment we assume that the valence and conduction bands have extrema at $\mathbf{k} = 0$ and are of the standard form

$$W_m = E_m - \hbar^2 k^2/2m_h^* \tag{4.24}$$

and

$$W_n = E_n + \hbar^2 k^2/2m_e^* \tag{4.25}$$

where m_h^* and m_e^* are considered positive. Then Eq. (4.23) becomes

$$\left[-\frac{\hbar^2}{2\mu}\nabla^2 - \frac{e^2}{\beta} - \frac{\hbar^2}{2i}\left(\frac{1}{m_h^*} - \frac{1}{m_e^*}\right)\mathbf{K}\cdot\nabla\right] U'$$

$$= \left[E - E_0^{(1)} - E_G - \frac{1}{8\mu}\hbar^2 K^2\right] U' \tag{4.26}$$

where $E_G = E_n - E_m$ is the energy gap and μ is the reduced mass defined in Eq. (4.7). This equation is precisely analogous to Eq. (4.6), and it follows that a transformation like (4.8) will eliminate the $\mathbf{K}\cdot\nabla$ term. Thus we set

$$U' = e^{i\alpha\mathbf{K}\cdot\mathbf{\beta}}\, F(\mathbf{\beta}) \tag{4.27}$$

and find that F must satisfy

$$\left(-\frac{\hbar^2}{2\mu}\nabla^2 - \frac{e^2}{\beta}\right)F = \left[E - E_0^{(1)} - E_G - \frac{\hbar^2 K^2}{2(m_e^* + m_h^*)}\right] F. \tag{4.28}$$

Hence the eigenvalues of $\mathcal{H}_{\beta\beta'}$ in the Wannier approximation are

$$E_{\nu\mathbf{K}}^{(0)} = E_0^{(1)} + E_G - \frac{\mu e^4}{2\hbar^2 n^2} + \frac{\hbar^2 K^2}{2(m_e^* + m_h^*)}. \tag{4.29}$$

$E_0^{(1)}$ is the first-order, ground-state energy, and $E_G - \mu e^4/2\hbar^2 n^2$ the additional energy required to create an electron-hole pair in a zero-order bound state $F_{nlm}(\beta)$; the set of quantum numbers nlm is denoted by ν. The last term in Eq. (4.29) is the kinetic energy of the bound pair. The zero-order states of the crystal corresponding to the eigenvalues [Eq. (4.29)] are [see (4.27), (4.18), and (2.31)]

$$\Psi_{\nu \mathbf{K}} = \sum_{\beta} U_{\nu \mathbf{K}}(\beta) \, \Phi(\mathbf{K}, \beta) = \sum_{\beta} e^{i\alpha' \mathbf{K} \cdot \beta} F_\nu(\beta) \, \Phi(\mathbf{K}, \beta) \qquad (4.30)$$

where $\alpha' = \tfrac{1}{2} + \alpha = m_e^*/(m_e^* + m_h^*)$. $|F_\nu(\beta)|^2$ is thus the probability that the system is in the state $\Phi(\mathbf{K}, \beta)$, i.e., the probability that the electron is at site β if the hole is at the origin. Although $F_\nu(\beta)$ as used in Eq. (4.30) is strictly defined as a point function, taking N values within the fundamental Born-von Karman parallelepiped of N atoms, it has been computed without regard to this restriction (the hydrogenic functions are defined and normalized over all space). This is justified in crystals having dimensions large compared with the exciton radii predicted (~ 100 A) and provided we normalize F such that $F_\nu(\beta)$, regarded as a pure number, satisfies

$$\sum_{\beta} |F_\nu(\beta)|^2 = 1. \qquad (4.31)$$

This is possible if $F_\nu(\beta)$ is taken as $\Omega_0^{1/2}$ times the average value of a hydrogenic wave function in the unit cell containing β. Here Ω_0 is the unit cell volume.

The eigenvalues (4.29) are called zero-order because $\mathscr{I}_{\beta\beta'}(\mathbf{K})$ has yet to be included. When it is included, we shall obtain first-order eigenvalues. No dielectric constant will appear in Eq. (4.29), because this is a result of yet higher-order effects, as discussed in Sections 4b and 7. Before discussing $\mathscr{I}_{\beta\beta'}(\mathbf{K})$, we sketch the derivation of zero-order states when the conduction and valence bands are not of standard form.

Degeneracies at conduction or valence band extrema will generally lead to anisotropic effective mass tensors, in which case a unique center-of-mass transformation makes it impossible to eliminate cross terms [such as the one involving $\mathbf{K} \cdot \nabla$ in Eq. (4.26)] by merely changing the phase of the wave function [as in (4.27)]. Dresselhaus points out,[20] however, that the $\mathbf{K} \cdot \nabla$ terms may in any case be treated as a perturbation. Suppose that $\mathbf{K} \cdot \nabla$ is neglected in Eq.

(4.26). Then U' itself is a hydrogenic function and the approximate eigenvalues are

$$E_0^{(1)} + E_G - \mu e^4/2\hbar^2 n^2 + \hbar^2 K^2/8\mu \qquad (4.32)$$

as compared with the true zero-order eigenvalues (4.29). The $\mathbf{K} \cdot \nabla$ term contributes a second-order energy

$$-\frac{\hbar^2}{4}\left(\frac{1}{m_e^*} - \frac{1}{m_h^*}\right)^2 \sum_{\nu' \neq \nu} \frac{|(U'_{\nu'}| \mathbf{K} \cdot \nabla | U'_{\nu})|^2}{E_{\nu'\mathbf{K}} - E_{\nu\mathbf{K}}} = -\frac{1}{8\mu}\left(\frac{1}{m_h^*} - \frac{1}{m_e^*}\right) \hbar^2 K^2 \qquad (4.33)$$

which, added to Eq. (4.32), changes the last term to $\hbar^2 K^2/2(m_e^* + m_h^*)$. A closed form is possible for the sum in Eq. (4.33) because of an f-sum rule for hydrogenic states.[20] In general, an f-sum rule is not as useful as it is in this case for effecting the second-order summation. In a considerably more general case, let the conduction and valence bands have the form

$$W_n(\mathbf{k}) = \mathcal{W}_n(\mathbf{k} - \mathbf{k}_{e0}) = E_n + \sum_{\mu,\nu} (\hbar^2/2m_e^*)_{\mu\nu} (\mathbf{k} - \mathbf{k}_{e0})_\mu (\mathbf{k} - \mathbf{k}_{e0})_\nu \qquad (4.34)$$

and

$$W_m(\mathbf{k}) = \mathcal{W}_m(\mathbf{k} - \mathbf{k}_{h0}) = E_m - \sum_{\mu,\nu} (\hbar^2/2m_h^*)_{\mu\nu} (\mathbf{k} - \mathbf{k}_{h0})_\mu (\mathbf{k} - \mathbf{k}_{h0})_\nu . \qquad (4.35)$$

Then the general Wannier equation, (4.23), becomes

$$[\mathcal{W}_n(-i\nabla + \tfrac{1}{2}\varkappa) - \mathcal{W}_m(-i\nabla - \tfrac{1}{2}\varkappa) - e^2/\beta] U'' = (E - E_0^{(1)}) U'' \qquad (4.36)$$

where $\varkappa = \mathbf{K} - \mathbf{k}_{e0} + \mathbf{k}_{h0}$ and $U'' = e^{-i(\mathbf{k}_{0h}+\mathbf{k}_{0e})\cdot\boldsymbol{\beta}/2} U'$. This may be written as a nearly-hydrogenic equation,

$$\left[-\frac{\hbar^2}{2}\sum_{\mu,\nu}\left(\frac{1}{m_e^*} + \frac{1}{m_h^*}\right)_{\mu\nu} \frac{\partial}{\partial \beta_\mu} \frac{\partial}{\partial \beta_\nu} - \frac{e^2}{\beta}\right] U''$$

$$= \left[E - E_0^{(1)} - E_G - \frac{\hbar^2}{2}\sum_{\mu,\nu}\left(\frac{1}{m_e^*} + \frac{1}{m_h^*}\right)_{\mu\nu} \varkappa_\mu \varkappa_\nu\right] U'' \qquad (4.37)$$

if the cross-terms

$$-\tfrac{1}{4} i\hbar^2 \sum_{\mu,\nu} \left(\varkappa_\mu \frac{\partial}{\partial \beta_\nu} + \varkappa_\nu \frac{\partial}{\partial \beta_\mu}\right)\left(\frac{1}{m_e^*} - \frac{1}{m_h^*}\right)_{\mu\nu} U'' \qquad (4.38)$$

are neglected.

Even if one were able to find the solutions of Eq. (4.37), it would not be an easy matter to compute corrections due to (4.38), thereby obtaining the zero-order Wannier effective mass. At $\varkappa = 0$, the solutions of Eq. (4.37) can be obtained by a variational method similar to the one employed by Kohn and Luttinger.[68] Detailed calculations based on Eq. (4.37) will, in fact, share all the features of the shallow impurity-state problem, such as the need to consider linear combinations of exciton states constructed from conduction and valence bands at several equivalent extrema in crystals, such as Ge, in which these occur. The only detailed calculations for exciton states along these variational lines of which the author is aware are those of Hopfield and Thomas[53] and Wheeler and Dimmock.[69]

Returning to the simpler case of standard valence and conduction bands, we compute the matrix of $\mathscr{H}_0$ in our new representation (4.30) in order to include first order effects. Using Eqs. (4.30) and (4.14), we have

$$(mn\nu \mathbf{K} \,|\, \mathscr{H}_0 \,|\, mn\nu' \mathbf{K}) = \sum_{\beta} \sum_{\beta'} U_{\nu \mathbf{K}}(\beta)^* \, U_{\nu' \mathbf{K}}(\beta') [\mathscr{H}_{\beta\beta'}(\mathbf{K}) + \mathscr{I}_{\beta\beta'}(\mathbf{K})]. \quad (4.39)$$

But we know that $\mathscr{H}_{\beta\beta'}$ is diagonal in this representation; therefore Eq. (4.39) is just $\delta_{\nu\nu'} E_{\nu \mathbf{K}}^{(0)}$ plus the following first-order correction, which is easily obtained from Eqs. (4.39), (4.30), and (4.16):

$$\mathscr{I}_{\nu\nu'}(\mathbf{K}) = -\left[(\nu\mathbf{O}\, m\mathbf{O} \,|\, g \,|\, \nu'\mathbf{O}\, m\mathbf{O}) - \sum_{\beta} F_\nu(\beta)^* \, F_{\nu'}(\beta)\, e^2/\beta\right]$$
$$+ 2\delta_M(\nu\mathbf{O}\, m\mathbf{O} \,|\, g \,|\, m\mathbf{O}\, \nu'\mathbf{O})$$
$$+ \sum_{\mathbf{R} \neq 0} e^{i\mathbf{K}\cdot\mathbf{R}} [2\delta_M(\nu\mathbf{R}\, m\mathbf{O} \,|\, g \,|\, m\mathbf{R}\, \nu'\mathbf{O}) - (\nu\mathbf{R}\, m\mathbf{O} \,|\, g \,|\, \nu'\mathbf{O}\, m\mathbf{R})].$$
$$(4.40)$$

By introducing the linear combinations

$$\psi_{\nu \mathbf{R}}(\mathbf{r}) = \sum_{\beta} e^{i\varkappa' \mathbf{K} \cdot \beta}\, F_\nu(\beta)\, a_{n(\mathbf{R}+\beta)}(\mathbf{r}) \quad (4.41)$$

which are abbreviated $\nu\mathbf{R}$ in the matrix elements in Eq. (4.40) and which act as "molecular orbitals" for the excited electron in the

---

[68] W. Kohn and J. M. Luttinger, *Phys. Rev.* **98**, 915 (1955); see also the review in this series by W. Kohn, *Solid State Phys.* **5**, 257 (1957).

[69] R. G. Wheeler and J. O. Dimmock, *Phys. Rev.* **125**, 1805 (1962).

crystal, it has been possible to put (4.40) into a form directly comparable with the formalism of the Frenkel exciton. The first term in Eq. (4.40) essentially corrects the electron-hole coulomb interaction for the finite size of the particles and for the symmetry of the crystal, and it is easily seen that contributions to $(\nu O\ mO\ |\ g\ |\ \nu' O\ mO)$ from large β are canceled by the second term in square brackets. The second term of Eq. (4.40) is the electron-hole exchange interaction, which appeared in the Frenkel exciton as part of $W_{ll'}$ [Eq. (3.8)]. The localized nature of the Wannier functions make the products $\psi^*_{\nu O} a_{m O}$ and $a^*_{m O}\psi_{\nu' O}$ effectively vanish outside the unit cell at the origin, e.g.,

$$\psi_{\nu O}(\mathbf{r})^* a_{m O}(\mathbf{r}) = \sum_{\beta}' e^{-i\alpha' \mathbf{K} \cdot \boldsymbol{\beta}}\, F_\nu(\boldsymbol{\beta})^*\, a_{n\beta}(\mathbf{r})^*\, a_{m O}(\mathbf{r})$$

$$\approx F_\nu(0)^*\, a_{n O}(\mathbf{r})^*\, a_{m O}(\mathbf{r}) \qquad (4.42)$$

so the exchange interaction may be approximated closely by

$$2\delta_M\, F_\nu(0)^*\, F_{\nu'}(0)\, (n O\ m O\ |\ g\ |\ m O\ n O). \qquad (4.43)$$

This "spin-spin" interaction, as it is sometimes called, is appreciable only in Wannier states having s-like envelope functions F_{ns}. Elliott[70] has estimated it for the $1s$ state in Cu_2O using known atomic exchange integrals in Cu^+. Its computed order of magnitude is 0.002 ev. Using approximate wave functions for Cu and O^{--}, Moskalenko and Tolpygo[71] estimated the coulomb correction and exchange energy in Cu_2O, obtaining a correction to the $1s$ and $2s$ levels of the order of 0.01 ev. It is not possible to determine the extent of the exchange contribution to this correction from their published results.

Finally, the last term in Eq. (4.40) is analogous to the third term of the Frenkel energy (3.7). Thus we see how the longitudinal-transverse splitting discussed in the Frenkel case appears explicitly in the Wannier case as a first-order perturbation on the hydrogenic levels.[72] It will occur in singlet states or other states having a singlet component

---

[70] R. J. Elliott, *Phys. Rev.* **124**, 340 (1961).
[71] S. A. Moskalenko and K. B. Tolpygo, *Zh. Eksperim. i Teor. Fiz.* **36**, 149 (1959); see *Soviet Phys. J.E.T.P.* (*English Transl.*) **9**, 103 (1959).
[72] The splitting has also been computed for Wannier excitons by E. I. Rashba, *Zh. Eksperim. i Teor. Fiz.* **36**, 1703 (1959); see *Soviet Phys. J.E.T.P.* (*English Transl.*) **9**, 1213 (1959).

whenever the transition densities $a_{m\mathbf{O}}^{*}\psi_{\nu'\mathbf{O}}$ and $\psi_{\nu\mathbf{R}}^{*}a_{m\mathbf{R}}$ have dipole moments,

$$\mu = \int a_{m\mathbf{O}}(\mathbf{r})^{*}\,\mathbf{r}\psi_{\nu\mathbf{O}}(\mathbf{r})\,d\mathbf{r} = \sum_{\boldsymbol{\beta}} e^{-i\alpha'\mathbf{K}\cdot\boldsymbol{\beta}}\,F_{\nu}(\boldsymbol{\beta})\int a_{m\mathbf{O}}(\mathbf{r})^{*}\,\mathbf{r}a_{n\boldsymbol{\beta}}(\mathbf{r})\,d\mathbf{r}. \quad (4.44)$$

Because of the localized nature of the functions a, the term $\boldsymbol{\beta} = \mathbf{O}$ contributes the largest amount to this sum. Thus

$$\mu \approx F_{\nu}(\mathbf{O}) \int a_{m\mathbf{O}}(\mathbf{r})^{*}\,\mathbf{r}a_{n\mathbf{O}}(\mathbf{r})\,d\mathbf{r}. \quad (4.45)$$

This will be nonzero when, for example, one band is p-like and the other s-like, provided F_{ν} is an s-like envelope function [$F_{ns}(0) \neq 0$]. If F is not s-like, then $F_{\nu}(0) = 0$, but there may still exist a very small dipole moment if bands m and n have appropriate symmetries. The Wannier functions associated with different lattice sites are orthogonal, to be sure, but the presence of $\mathbf{r}$ makes possible a small contribution from nearest neighbors in the sum. An *appreciable* dipole moment exists, however, only when F is s-like, and the relative symmetries of a_m and a_n are appropriate. Only under these conditions will there be an appreciable longitudinal-transverse splitting. This will be the case whenever the exciton state $\Psi_{\nu\mathbf{K}}$ can be created by an "allowed" transition (Section 9).

Some $\mathbf{K}$ dependence of the exciton energy will enter in the various contributions to the first-order energy, particularly the one discussed last. This qualifies Dresselhaus' observation[20] that all Wannier excitons have the same mass independent of their hydrogenic state, a statement based on consideration of only the zero-order energy $E_{\nu\mathbf{K}}^{(0)}$, Eq. (4.29).

The question of whether an appreciable dipole moment exists in $a_{m\mathbf{O}}^{*}\psi_{\nu'\mathbf{O}}$ or $\psi_{\nu\mathbf{R}}^{*}a_{m\mathbf{R}}$ brings us naturally to a discussion of the over-all symmetry of zero-order Wannier exciton functions. Consider, first, the case $\mathbf{K} = 0$, for which Eq. (4.30) becomes

$$\Psi_{\nu 0} = \sum_{\boldsymbol{\beta}} F_{\nu}(\boldsymbol{\beta})\,\Phi(0, \boldsymbol{\beta}). \quad (4.46)$$

Each determinantal wave function in $\Phi(0, \boldsymbol{\beta})$ has the rotational symmetry of $\bar{a}_{m\mathbf{O}}a_{n\boldsymbol{\beta}}$, where $\bar{a}_{m\mathbf{O}}$ is a hole wave function in the sense that it duplicates the symmetry of the state created by removing an electron formerly in a_{m0}. Thus the rotational symmetry of $\Psi_{\nu 0}$ is the rotational symmetry of

$$\sum_\beta F_\nu(\beta)\, \bar{a}_{m0}\, a_{n\beta} = \bar{a}_{m0}\, \psi_{\nu 0}. \tag{4.47}$$

The results of group-theoretical treatments of exciton symmetries[73,74] at $\mathbf{K} = 0$ can be stated in the following form: the "angular momenta" of the hole, the electron, and the envelope function are added together and the various resultants are the possible "angular momenta" of the entire exciton state. In Eq. (4.47) we have suggested that the envelope function and the electron be combined first, and then coupled to the hole. In Fig. 10 we show some simple examples of how an exciton of overall P symmetry might be obtained. Here Fig. 10a shows an s-like envelope of s functions ($\psi_{\nu 0}$ is S-like), which, when combined with a p-like hole, gives an over-all P state. In Fig. 10b, we have an s-like envelope of p functions ($\psi_{\nu 0}$ is P-like), which, when combined with an s-like hole, gives again an over-all P state. In Fig. 10c, a p-like envelope of s functions ($\psi_{\nu 0}$ is P-like, but only because of the envelope this time) combines with an s-like hole.

As is well-known, atomic quantum numbers are of limited usefulness in most cases in solids. The "angular momentum addition," which might be written

$$p \text{ (hole)} + s \text{ (env)} + s \text{ (electron)} = P \text{ (exciton)}$$

in the case of Fig. 10a, is written in general group-theoretical terms as

$$\Gamma_m \text{ (hole)} \times \Gamma_\nu \text{ (env)} \times \Gamma_n \text{ (electron)} = \sum_\alpha c_\alpha \Gamma_\alpha \text{ (exciton)} \tag{4.48}$$

which means that when electrons and holes of the given symmetry types are coupled via a given envelope function symmetry, one can obtain c_1 sets of exciton states of type Γ_1, c_2 of type Γ_2, etc. The quantum numbers Γ_m and Γ_n appearing in Eq. (4.48) can refer to either the point symmetry of the Wannier functions in bands m and n or, if the extrema of the bands are at $\mathbf{k} = 0$, the symmetries of the corresponding Bloch functions at this point.[73]

When $\mathbf{K} \neq 0$ it is too difficult to write a general prescription

[73] J. J. Hopfield, *Phys. Chem. Solids* **15**, 97 (1960).
[74] M. Balkanski and J. des Cloizeaux, *J. Phys. Radium* **22**, 41 (1961).

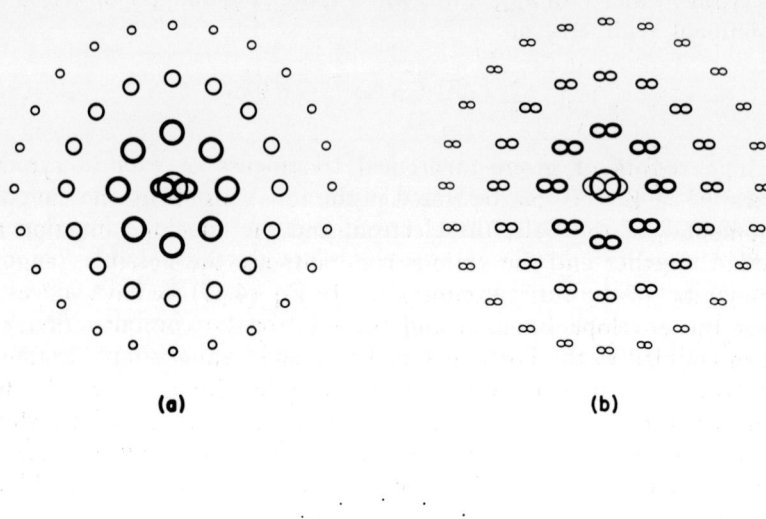

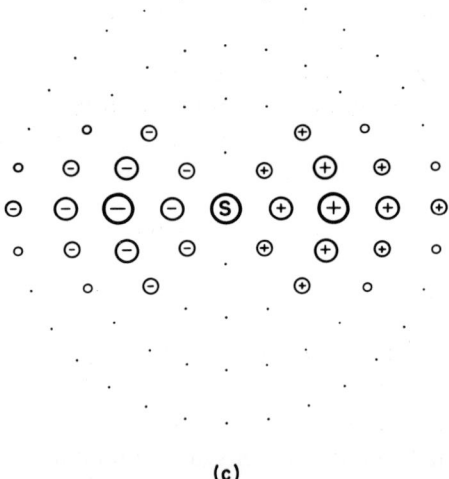

Fig. 10. Schematic illustration of certain Wannier exciton states with over-all P-like symmetry. On reflection in the vertical, each changes sign. The relative size of a localized orbital indicates roughly the relative probability of the electron being found there, if the hole is taken to be at the origin. "+" and "—" show relative phases. States shown in (a) and (b) are s-like in the hydrogenic model, but acquire over-all P character because of the hole in (a) or the electron in (b). The state shown in (c) is p-like in the hydrogenic model to begin with.

analogous to Eq. (4.48), but it seems clear that if states of known symmetry have been constructed at $\mathbf{K} = 0$, states in the vicinity may be obtained by inspection, with the aid of compatibility relations.[52]

b. The Dielectric Constant

In Section 4a we applied the general formalism of Section 2 and obtained the zero-order energy levels and wave functions for the Wannier exciton [Eq. (4.29)]. The dielectric constant entered nowhere, although in the simple two-particle model we had expected it to reduce the binding energy by a factor of ϵ^2. The present section gives a qualitative discussion of the dielectric constant in order to hasten to a comparison of the Wannier theory with experiment. Mott[9] appears to have been the first to suggest modifying the Wannier model to account for the dielectric medium.

The potential energy seen by the electron as it recedes from the hole will depend on how well it is polarizing the lattice. Suppose the electron is very close to the hole. The pair then has a very high internal kinetic energy and the remaining valence electrons cannot follow its internal motion if its frequency of rotation exceeds their effective resonant frequency, which may be approximated by $E_G/\hbar$. Thus for an exciton of radius β, we expect a simple Coulomb law to hold if its angular frequency $\omega = $ (angular momentum) $\times$ (mass $\times$ radius^2)$^{-1} \sim \hbar/\mu\beta^2$ exceeds $E_G/\hbar$, i.e., if $\beta \gtrsim (\hbar^2/\mu E_G)^{1/2} \equiv \beta_E$. For an exciton of mass $\mu \sim \frac{1}{2}m$ and in a solid in which $E_G \sim 2$ ev, $\beta_E \approx 5$ Bohr radii. This corresponds to an exciton of atomic dimensions. For $\beta \gtrsim (\hbar^2/\mu E_G)^{1/2}$, the relative motion of the electron and hole becomes slow enough that the valence electrons can follow them, and the ensuing polarization is expected to reduce the Coulomb law to $- e^2/\epsilon_\infty \beta$, where ϵ_∞ is roughly the square of the refractive index. As the radius becomes larger and larger, the frequency of the internal motion finally becomes so small that entire atoms can even follow it. This happens when $\omega \sim \hbar/\mu\beta^2$ is of the order of optical vibrational frequencies, i.e.,

$$\hbar/\mu\beta^2 \sim \omega_O \tag{4.49}$$

where $\omega_O \sim 3 \times 10^{13}$ sec^{-1}. Using this criterion, we find that a low-frequency dielectric constant should be used if

$$\beta \gtrsim \sqrt{\hbar/\mu\omega_0} \equiv \beta_L. \tag{4.50}$$

Again, for an exciton of mass $\mu \sim \frac{1}{2}m$, we have $\beta_L \approx 56 a_0$.

In Fig. 11 is sketched a rough summary of the preceding discussion, showing the probable transitions between the various regions of validity of the different Coulomb interactions. In practice, the

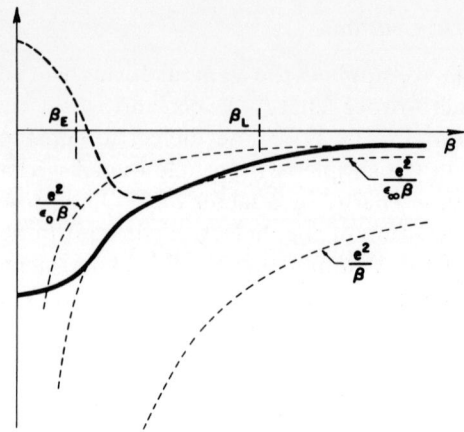

FIG. 11. Effective electron-hole interaction. For $\beta < \beta_E$ it becomes non-Coulombic because of the finite size of the electron and hole, and because of various exchange-like potentials. At $\beta > \beta_E$, the potential is likely to follow some form of Coulomb's law (see text).

effective potential never reaches $-e^2/\beta$ at small β, because of the short-range, first-order energy corrections. The region of space inside β_E is small relative to the volume occupied by the wave function of relative motion, however, so it is perfectly reasonable to assume that the interaction is $-e^2/\epsilon_\infty \beta$ in this region when obtaining zero-order wave functions. The transition from $-e^2/\beta$ to $-e^2/\epsilon_\infty \beta$ at small exciton radii has been studied by Haken and Schottky[75] and by Englert.[76]

A detailed investigation of the region $\beta \geqslant \beta_E$ by Haken (Section 7) indicates that the effective Coulomb interaction is given by

$$V(\beta) = -\frac{1}{\epsilon(\beta)} \frac{e^2}{\beta} \tag{4.51}$$

where

$$\frac{1}{\epsilon(\beta)} = \left[\frac{1}{\epsilon_\infty} - \left(\frac{1}{\epsilon_\infty} - \frac{1}{\epsilon_0}\right)\left(1 - \frac{e^{-\beta/\rho_e} + e^{-\beta/\rho_h}}{2}\right)\right] \tag{4.52}$$

[75] H. Haken and W. Schottky, Z. Physik. Chem. (Frankfurt) 16, 218 (1958).
[76] F. Englert, Phys. Chem. Solids 11, 78 (1959).

in which

$$\rho_e = \left(\frac{\hbar}{2m_e^*\omega_{LO}}\right)^{1/2} \quad \text{and} \quad \rho_h = \left(\frac{\hbar}{2m_h^*\omega_{LO}}\right)^{1/2};$$

ω_{LO} is the LO (longitudinal optical) phonon frequency. Thus $\epsilon(\beta)$ approaches ϵ_∞ and ϵ_0 at small and large β, respectively. The transformation from ϵ_∞ to ϵ_0 is essentially complete at the larger of ρ_e and ρ_h; in AgCl, for example, taking $m_e^* \sim 0.5m \ll m_h^*$, and $\omega_{LO} = 3.4 \times 10^{13}$ sec^{-1}, we find that a static dielectric constant is appropriate for exciton radii greater than $\beta_L \sim 35a_0$.

A solution of the Schrödinger equation

$$\left(-\frac{\hbar^2 \nabla^2}{2\mu} - \frac{1}{\epsilon(\beta)}\frac{e^2}{\beta}\right) F = EF \tag{4.53}$$

with $\epsilon(\beta)$ given by Eq. (4.52) has not been attempted because of its analytic complexity. Hrivnak[77] has solved it for the special case $\epsilon_0 = 2\epsilon_\infty$ and $m_e^* = m_h^*$, but only for s states of low energy. Not surprisingly, he obtains nearly hydrogenic wave functions with damping lengths slightly larger than those in the case $\epsilon_0 = \epsilon_\infty$, accompanied by slightly higher energy levels. Both of these can be attributed to the weakened tail of the Coulomb interaction. It is clear from the details and restrictions of his calculation that useful solutions of Eq. (4.53) will probably best be found numerically. One qualitative statement which may be made with no computation at all is that the eigenvalues of Eq. (4.53) will no longer necessarily have Coulomb degeneracy, e.g., F_{2s} and F_{2p} may well lie at different energies. This can also result from the correction terms which we have earlier called "first-order". Moskalenko and Tolpygo[71] compute an approximate $2s - 2p$ splitting of 0.0005 ev in Cu$_2$O on the basis of the first-order terms.

The traditional method of handling the problem of the dielectric constant has been to fit the hydrogenic eigenvalues $-\mu e^4/2\hbar^2\epsilon^2 n^2$ to observed energy levels, thus determining the exciton rydberg experimentally. A choice of ϵ then determines μ, which can be checked with other data, or, the spatial distribution of the wave functions associated with the value of ϵ chosen can be compared with the valid ranges of ϵ. In Section 4c we will look into a few actual cases.

[77] L. Hrivnak, *Czech. J. Phys.* **9**, 685 (1959).

The variation of the dielectric constant with direction has been accounted for by Barriol et al.[78] and by Hopfield and Thomas[53] for the case of uniaxial crystals. The two calculations, both essentially perturbation treatments of the Wannier equation, involve different anisotropy parameters and are not readily compared. The latter work provides for mass anisotropy and is thus more generally useful.

c. Comparison with Experiment

There are many features of the Wannier exciton model which are subject to experimental test, but certainly the most striking is the hydrogen-like spectrum derived in the preceding section. Optical absorption or reflection is used to locate exciton levels; this will be studied in detail in Sections 8-10. At the moment we will merely continue to assume that with some care, one can interpret energies of absorption peaks as energy levels of the crystal relative to its ground state. We might ask which crystals appear to display a hydrogenic series of levels on the basis of the observed lines. There is actually a small number, if we admit only those cases in which at least three levels follow the scheme $E = E_G - \mu e^4/2\hbar^2 n^2 \epsilon^2$ (excluding the $n = 1$ level, because it has a small radius and is sensitive to first-order perturbations).

(1) *Cuprous Oxide, Cu_2O.*[79] Several hydrogenic series appear in this crystal. At 4.2° K absorption lines of a "yellow" series appear at energies

$$E = 2.1726 - (0.0968/n^2) \text{ ev} \qquad (n = 2, \cdots, 6) \qquad (4.54)$$

and those of a "green" series at

$$E = 2.3058 - (0.1540/n^2) \text{ ev} \qquad (n = 2, 3, 4). \qquad (4.55)$$

Observed and fitted energy levels are given in Table I, and recent data by Baumeister[80] illustrating the yellow series were shown in Fig. 2. Though we have used Nikitine's 4.2° K data[17] in the table, it should be noted that Gross[16] has managed to identify lines in the

---

[78] J. Barriol, S. Nikitine, and M. Sieskind, *Compt. Rend.* **242**, 790 (1956).
[79] M. Hayashi and K. Katsuki, *J. Phys. Soc. Japan* **5**, 381 (1950); **7**, 599 (1952); E. F. Gross and N. A. Karryeff, *Dokl. Akad. Nauk S.S.S.R.* **84**, 261, 471 (1952).
[80] P. W. Baumeister, *Phys. Rev.* **121**, 359 (1961).

series as high as $n = 9$ and 10 at $1.3°$ K! Furthermore, Gross and Kuang-Yin[81] have located hydrogenic series in the blue and violet, presumably associated with pairs of valence and conduction bands separated by band gaps much larger than 2.3 ev.

TABLE I. OBSERVED AND CALCULATED WANNIER EXCITON ENERGY LEVELS IN Cu_2O AT $4.2°$ K

"Calculated" refers to fitting of the levels by Eqs. (4.55) and (4.56). λ and E are wavelength in A and energy in ev, respectively.

Band (n)	"Yellow series"			"Green series"		
	$\lambda_{obs.}$	$E_{obs.}$	$E_{calc.}$	$\lambda_{obs.}$	$E_{obs.}$	$E_{calc.}$
∞	5707	2.1724	2.1726	5377	2.3057	2.3058
6	5713.2	2.1700	2.1699	—	—	2.3015
5	5716	2.1690	2.1687	—	—	2.2996
4	5722.4	2.1665	2.1665	5401.9	2.2951	2.2962
3	5735.2	2.1617	2.1618	5417.1	2.2887	2.2887
2	5770.6	2.1485	2.1484	5468.1	2.2673	2.2673
1	6125	2.032	2.0758	—	—	2.1518

The states corresponding to the eigenvalues (4.54) and (4.55) have symmetries such that optical absorption in the corresponding lines is quite weak, and the line $n = 1$ is particularly weak, having zero oscillator strength in a first approximation (Section 9). Nevertheless, the yellow-series $n = 1$ line is observed[82] at 2.032 ev, or 0.044 ev below the energy predicted by (4.54). No $n = 1$ level of the green series has been located, but one would expect it to lie at ~ 2.11 ev.

Because of some uncertainty in the dielectric constants of Cu_2O, it is hard to evaluate the reduced mass with any precision. Haken[83] estimates that ϵ_0 is appropriate to all but the $n = 1$ state, which is certainly borne out by the yellow series data. Using information obtained[82] from anisotropic absorption in the $n = 1$ line, Elliott[70] has proposed a tentative band structure of Cu_2O near $k = 0$, incorporating his earlier suggestion[21] that the existence of two series is due

[81] E. F. Gross and C. Kuang-Yin, *Fiz. Tverd. Tela* **4**, 261 (1962); see *Soviet Phys.—Solid State (English Transl.)* **4**, 186 (1962).

[82] E. F. Gross and A. A. Kaplianskii, *Dokl. Akad. Nauk S.S.S.R.* **132**, 98 (1960); *Soviet Phys.—Solid State* **2**, 353 (1960).

[83] H. Haken, in "Halbleiterprobleme" (W. Schottky, ed.), Vol. IV, p. 1. Vieweg, Braunschweig, 1958.

to the presence of two valence bands, each cooperating with the conduction band to produce a set of exciton states. We have sketched this structure in Fig. 12. From the observed exciton rydbergs [Eqs. (4.54), (4.55)] of 0.0968 and 0.1540 ev, we compute exciton reduced masses of $0.00711\ \epsilon_0^2 m$ and $0.0113 \epsilon_0^2 m$, respectively, for the yellow and green series.

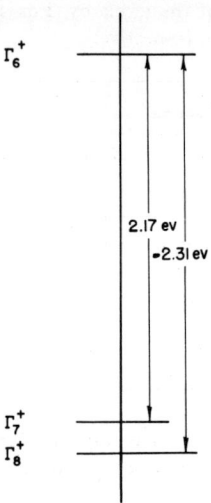

FIG. 12. Tentative band structure of Cu_2O at $\mathbf{k} = 0$ as proposed by R. J. Elliott [*Phys. Rev.* **124**, 340 (1961)]. The notation Γ_i^+ indicates even-parity representations of the cubic double group [see, e.g., G. F. Koster, *Solid State Phys.* **5**, 173 (1957)].

Elliott's interpretation of the existence of two hydrogenic series given here is only one of three alternative proposals, but the one we consider most viable. Nikitine has suggested that the series are "vibrational satellites" of one another and thus correspond to the same electronic states. This proposal was based in part on the existence of a third "red" series which lay at the position of the yellow series $n = 1$ line and may have been confused with it. Pavinskii and Zhilich[84] proposed the existence of two hole bands, odd and even under inversion, which would amount in a sense to a crystal splitting of (say) a Cu^+ 3d band. Undoubtedly both spin-orbit splitting and crystal

---

[84] P. P. Pavinskii and A. G. Zhilich, *Vestn. Leningr.* **12**, No. 4, *Ser. Fiz. i Khim.* No. 1, 50 (1957).

splitting play some role in setting up two valence bands, but the close agreement between the spin-orbit parameter in Cu^{++} and the observed splitting make it seem reasonable to accept Elliott's proposal.

(2) *Cadmium Sulfide, CdS.* Early work by Gross' group[85] indicated a possible hydrogenic series in CdS, and Dutton[86] discovered dichroic effects in reflectivity which were expected to be associated with intrinsic absorption peaks. Thomas and Hopfield, in an extensive study of thin crystals,[49,53,87–89] have been able to arrive at a consistent picture of the CdS band structure at the valence and conduction band various conditions. Their findings are sketched in Fig. 13. Associated with the three valence bands shown there are three exciton

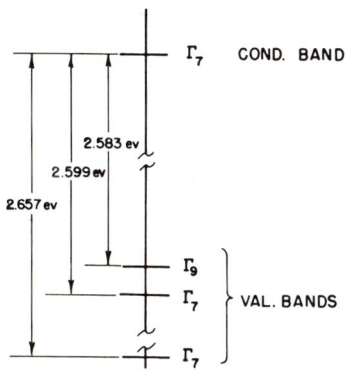

FIG. 13. Tentative band structure of CdS at $\mathbf{k} = 0$ as deduced by D. G. Thomas and J. J. Hopfield from studies on exciton spectra. Notation is that of H. A. Bethe [*Ann. Phys.* **3**, 133 (1929)] for the hexagonal group (C_{6v}); the conduction band is s-like and the valence band is p-like, split by the crystal field and spin-orbit coupling.

series, denoted by A, B, and C in order of increasing gap energy; series A is constructed from a Γ_7 electron and a Γ_9 hole, and will claim our attention here because its members $n = 1, 2, 3,$ and 4 have been indentified. CdS is anisotropic, but the anisotropy is small

---

[85] E. F. Gross, B. S. Razbirin, and M. A. Jakobsen, *Zh. Tekhn. Fiz.* **27**, 1149 (1957), see *Soviet Phys.—Tech. Phys. (English Transl.)* **2**, 1038 (1957).
[86] D. B. Dutton, *Phys. Rev.* **112**, 785 (1958).
[87] D. G. Thomas and J. J. Hopfield, *Phys. Rev.* **116**, 573 (1959).
[88] D. G. Thomas, J. J. Hopfield, and M. Power, *Phys. Rev.* **119**, 570 (1960).
[89] D. G. Thomas and J. J. Hopfield, *Phys. Rev. Letters* **5**, 505 (1960); *Phys. Rev.* **124**, 657 (1961).

enough[53] that its effect on the energy can be regarded as a first-order perturbation; it mainly lifts the hydrogenic l and m degeneracies. The experimental and theoretical A-series energies are listed in Table II.

TABLE II

TABLE II. Observed and Calculated Wannier Exciton Levels Arising from the Lowest Conduction Band and Highest Valence Band in CdS
(after Hopfield and Thomas[53])

Energies are in ev and wavelengths in A. The structure of the $1S$ level is discussed in the text. λ_{refl} denotes wavelengths of lines seen in the corresponding reflection spectrum (Dutton[86]).

Band (n)	States (m)	$E_{\text{calc.}}$	$E_{\text{obs.}}$	$\lambda_{\text{obs.}}$	$\lambda_{\text{refl.}}$
∞	—	2.5826	—	—	—
4	$P_{\pm 1}$	2.5810			
	$D_{\pm 2}$	2.5809	2.58094	4803.28	—
	$D_{\pm 1}$	2.5809			
3	$P_{\pm 1}$	2.5797	—	—	—
	$D_{\pm 2}$	2.5797	2.57977	4805.46	—
	$D_{\pm 1}$	2.5796	—	—	—
	P_0	2.5794	—	—	—
2	$P_{\pm 1}$	2.5761	2.57575	4812.96	4844
	S	2.5759			
	P_0	2.5755	2.57508	4814.21	—
1	$S(L)$		2.5546	4852.9	—
	$S(T)$	2.5556 (av.)	2.5537	4854.6	4874
	$S(\Gamma_6)$		2.5524	4857.0	—

Hopfield and Thomas[53] should be consulted for details of the various effective masses. We have included some additional structure in the $1S$ levels in Table II in order to illustrate a complexity which pervades the presently known CdS spectrum, and which can be expected to occur in many crystals as resolution is increased and crystal perfection improved. Recall that $1S$ refers to the hydrogenic envelope function; different couplings of Γ_7 and Γ_9 according to this envelope give rise to states of *over-all* symmetry Γ_5 (P_x- and P_y-like) and Γ_6 (no simple analog). The Γ_6 level is observed only as a very weak absorption line, but the Γ_5 states, being P-like, have transition dipole moments and give rise to strong absorption lines. Also, their energies are affected by the long-range interaction discussed earlier. If the exciton **K** vector lies along the z direction (c axis), P_x and P_y are equivalent and one can construct two degenerate transverse excitons. If, however, **K** lies in the xy plane, one of the P states must be chosen to lie longitudinally with respect to **K** and the other transverse to it. The longitudinal-transverse splitting is seen in the CdS data, as denoted in the $1S$ case by L and T. (Whenever **K** is not precisely parallel to the xy plane, some transverse states become mixed with the longitudinal state. This, as we will see quantitatively in Section 9, enables the longitudinal state to be observed as an essentially allowed absorption line. As the **K** vector is tilted farther away from the xy plane this state assumes more and more transverse character, until it is completely transverse when **K** $\parallel c$.) The CdS "B series," formed from the conduction band and middle valence band, has members at 2.5687 ev ($n = 1$) and 2.5909 ev ($n = 2$).

(3) *Lead Iodide, PbI_2.* An intense hydrogenic series is observed by French workers[90] in PbI_2. Four lines can reasonably be identified and matched with a hydrogenic formula

$$E = 2.5677 - (0.1418/n^2) \text{ ev} \tag{4.56}$$

except, as usual, for the $n = 1$ line, which in this case lies too *high* by 0.08 ev. PbI_2 has hexagonal structure and the absorption spectrum shown corresponds to light propagating along the c axis. A somewhat similar spectrum is found in HgI_2, but there are not enough reasonable absorption lines to meet our criteria. The exciton rydberg of 0.142 ev in PbI_2 corresponds to a reduced mass of $\mu = 0.010 \times \epsilon_0^2 m$. Unfortunately, Imai[91] has recently been unable to locate this particular hydrogenic series.

(4) *Silver Iodide, γ-AgI.* In the γ phase of AgI, part of a hydrogenic series is observed,[92] with

$$E = 2.987 - (0.222/n^2) \text{ ev}. \tag{4.57}$$

The lines $n = 1, 2$ are obscured by a broad absorption band of presently unknown origin, if they exist at all. If the γ-AgI lines correspond to exciton absorption, the rydberg of 0.222 ev means a reduced mass of 0.016 $\epsilon_0^2 m$.

[90] S. Nikitine, J. Burckel, J. Biellman, and R. Reiss, *Compt. Rend.* **251**, 935 (1960).
[91] I. Imai, *Phys. Chem. Solids* **22**, 81 (1961).
[92] G. Perny, *J. Chim. Phys.* **55**, 650 (1958).

(5) *Other Crystals.* Although the emphasis of this section has been mainly on energy level schemes, there are clear cases of applicability of the Wannier model in the absence of a tell-tale spectrum. Cases in which only a few low-lying levels are observable include ZnO,[93] CdSe,[94,95] alkali azides,[96] and Ge.[97,98] In Ge, the exciton at $\mathbf{K} = 0$ has a tiny rydberg of about 0.0015 ev, and its series is blurred in absorption. In certain crystals (Si,[98] SiC,[99] AgCl,[100] and again Ge) "indirect excitons," meaning excitons formed at points in the BZ other than the center, are found. Although these indirect excitons presumably have just as respectable an energy-level scheme as the direct excitons (see Section 4a), sharp absorption lines are not seen because phonons of various wave vectors and energies assist in their actual formation. This topic is handled in Section 10c. Further observations related to Wannier excitons have been made on GaAs,[101] GaSb,[102] and ZnSe,[103] and other measurements will be mentioned in later sections where appropriate.

d. *Range of Validity and Usefulness of the Wannier Model*

The Wannier model is not appropriate for small-radius excitons for reasons other than the fact that first-order corrections are large and hard to calculate. Implicit in our derivation was the assumption that the valence and conduction bands were parabolic. This is certainly not true over the whole Brillouin zone (BZ), but it does not matter if the exciton has a large radius; the uncertainty principle then

---

[93] D. G. Thomas, *Phys. Chem. Solids* **15**, 86 (1960); R. E. Dietz, J. J. Hopfield, and D. G. Thomas, *J. Appl. Phys.* **32**, 2282 (1961).

[94] E. F. Gross and V. V. Sobolev, *Fiz. Tverd. Tela* **2**, 406 (1960); see *Soviet Phys.— Solid State (English Transl.)* **2**, 379 (1960).

[95] R. G. Wheeler and J. O. Dimmock, *Phys. Rev.* **125**, 1805 (1962).

[96] S. K. Deb, *J. Chem. Phys.* **35**, 2122 (1961).

[97] G. G. Macfarlane, T. P. McLean, J. E. Quarrington, and V. Roberts, *Phys. Rev.* **108**, 1377 (1957); *Proc. Phys. Soc. (London)* **71**, 863 (1958).

[98] G. G. Macfarlane, T. P. McLean, J. E. Quarrington, and V. Roberts, *Phys. Rev.* **111**, 1245 (1958); *Phys. Chem. Solids* **8**, 388 (1959).

[99] W. Choyke and L. Patrick, *Proc. Intern. Conf. Semicond. Phys., Prague, 1960,* p. 432 (1961).

[100] F. C. Brown, T. Masumi, and H. H. Tippins, *Phys. Chem. Solids* **22**, 101 (1961).

[101] M. D. Sturge, *Phys. Rev.* **128**, 768 (1962).

[102] E. J. Johnson and H. Y. Fan, *Bull. Am. Phys. Soc.* [2] **7**, 185 (1962).

[103] E. F. Gross, L. G. Suslina, and P. A. Kon'kov, *Fiz. Tverd. Tela* **4**, 396 (1962); see *Soviet Phys.—Solid State (English Transl.)* **4**, 287 (1962).

guarantees that only a small region of **k** space need be used to construct the wave functions. On the other hand, if the electron and hole are closely associated (β small), it takes a correspondingly large segment of the BZ to describe them. Therefore, although the transcription from a difference to a differential equation may be justified, cutting off the expansion of the band energies at terms of order k^2 is not; a proper treatment would result in a differential equation of very high order.

If the bands are parabolic in **k** space over a region of linear dimensions of the order of $\mu/m\epsilon a_0$, i.e., the reciprocal "exciton Bohr radius," then even the hydrogenic ground state will not be appreciably in error because of the cutoff at powers of k^2 (see the discussion of the analogous situation in shallow impurity states by Kohn,[68] p. 280ff.) Thus in Cu_2O, the Wannier differential equation is good even for a zero-order "yellow series" hydrogenic $1s$ function if the valence and conduction bands are parabolic out to 0.11×10^8 cm^{-1}, which is probably the case, since this involves only 15 % of the distance from the center of the BZ to the boundary in the (100) direction. It is clear that in crystals with small static dielectric constants, these conditions will not obtain. Moskalenko and Tolpygo[71] have actually included terms of order $(-i\nabla)^4$ as a perturbation on the exciton levels in Cu_2O, only to find that the first-order corrections, mentioned earlier, were partly canceled, bringing their final results very close to those of the simple hydrogenic model.

For excitons of extremely large radii, the Wannier model is most certainly valid, but is less useful because the orbits may attain dimensions comparable with those of the crystal. The excitons will then interact strongly with the surface, or, in other words, running exciton states will not be useful approximations to stationary states of the static crystal. In summary, the Wannier model is valid for most low-lying exciton states expected to exist in crystals with fairly large dielectric constants, and is probably valid though not as useful for higher bound states in all crystals.

5. Intermediate Cases

The solids whose fundamental optical absorption spectra were studied earliest, namely, the alkali halides and related crystals, remain today the least understood with respect to a detailed knowledge of the electronic structure of their lowest excited states. Suffice it to say that the earliest (1927) formula for the excitation energy in these crystals, which is demonstrably incorrect for the model used, still gives the most precise means of computing this energy! This section is devoted to a study of the models introduced in an attempt to bridge the gap between large- and small-radius excitons. Certain of the alkali halides will be used as prototypes in our discussion.

a. Wave Functions and Eigenvalues

Within their usual contexts, the Frenkel and Wannier models are both inappropriate to describe the lowest exciton states in the alkali halides. The Frenkel model in its strict sense is based on excited states of individual atoms. In the ground electronic state of the alkali halides, each alkali atom has lost an electron to a halogen atom, and it is these extra electrons on the halide sublattice which are expected to be easiest to excite. But there exists no bound excited state of a negative halogen ion, so the Frenkel model cannot apply formally. The closest substitute to "excited halide functions" would be excited states computed *in situ*, i.e., computed as solutions of the problem of a localized excitation in the actual crystal. It is clear from the results on argon (Section 3d) that such a calculation would be quite difficult because of large overlapping of the excited state wave function with its neighbors and the concomitant need to consider the details of the potentials of several neighbors, which include here both positive and negative ions. In the alkali halides one encounters just as much trouble on the Wannier model. Because of relatively small dielectric constants the effective Bohr radius is small and Bloch functions from quite a large portion of the BZ must be used to make up the exciton state. For example, in KCl, generously assuming that the static dielectric constant $\epsilon_0 = 4.7$ is effective for the 1s state, and that $\mu \sim \tfrac{1}{2} m$, we find that the reciprocal effective Bohr radius is $0.25 k_{ZB}$ where k_{ZB} is the value of k at the (100) BZ boundary.

Whatever the detailed form of the charge distribution within the exciton, certain general statements may be made on the basis of the results of the preceding sections. The eigenvalues will consist of a constant term, analogous to W_{ll} and $E_G - \mu e^4/2\hbar^2 \epsilon^2 n^2$ in the Frenkel and Wannier cases, respectively, which represents the energy required to create the pair locally; and a **K**-dependent term associated with the motion of the pair through the lattice. Associated with this latter term will be a longitudinal-transverse splitting under appropriate conditions.

Few attempts have been made to bridge the gap between the Frenkel and Wannier limits rigorously. Takeuti[104] cast the general secular Eq. (2.31) into a form involving a Green's function, analogous to that which Lax[105] and Koster and Slater[106] derived in connection with lattice vibrations and impurity states, with the result

$$U(\beta') = \sum_{\beta} \frac{1}{N} \sum_{k} \frac{e^{-i\mathbf{k}\cdot(\beta-\beta')}}{E - [W_n(\mathbf{k}) - W_m(\mathbf{k} - \mathbf{K})]} V_{\beta\beta}(\mathbf{K}) U(\beta) \quad (5.1)$$

where $V_{\beta\beta}$ is the "perturbation" on the total Bloch Hamiltonian, i.e., it includes both $\mathscr{I}_{\beta\beta}$ and $-e^2/\beta$ [cf. Eqs. (4.15) and (4.16)]. Unfortunately, the relative simplicity of Eq. (5.1) is a result of an assumption that V is diagonal in β, which is probably good only in the limit of large $|\beta|$, where the solution is already known. Furthermore, to obtain quantitative results Takeuti had to assume parabolic bands and restrict nonzero values of V to a very small range of β. Since V contains $-e^2/\beta$, this invalidated his results at the Wannier limit, and he obtained only the correct Wannier translational effective mass, not the hydrogenic spectrum. In spite of these limitations, we feel that the Green's function approach can be fruitfully generalized to handle both limits properly, perhaps by including $-e^2/\beta$ in the unperturbed Hamiltonian, in which case $V_{\beta\beta}$ will be replaced by only the truly short-range term $\mathscr{I}_{\beta\beta}$. A transcription to the $\Psi_{\nu\mathbf{K}}$ representation might facilitate the inclusion of non-diagonal terms of $\mathscr{I}_{\beta\beta'}$.

Most theoretical work on excitons in the alkali halides has been concerned with more direct and intuitive models of the excitons' internal electronic structure, i.e., the details of the electron and hole charge distributions in low-lying states. In Sections 5b and 5c we describe work along the lines of the "transfer" and "excitation" models, respectively.

b. The Electron-Transfer Model

As a result of their early investigations of alkali-halide absorption

---

[104] Y. Takeuti, *Progr. Theoret. Phys. (Kyoto)* **18**, 421 (1957); *Progr. Theoret. Phys. (Kyoto) Suppl.* **12**, 75 (1959).
[105] M. Lax, *Phys. Rev.* **94**, 1391 (1954).
[106] G. F. Koster and J. C. Slater, *Phys. Rev.* **95**, 1167 (1954).

spectra, Hilsch and Pohl[6] speculated that the primary effect of the absorption of a photon was to transfer an electron from one of the halogen atoms to a neighboring alkali atom (Fig. 14a). This model was

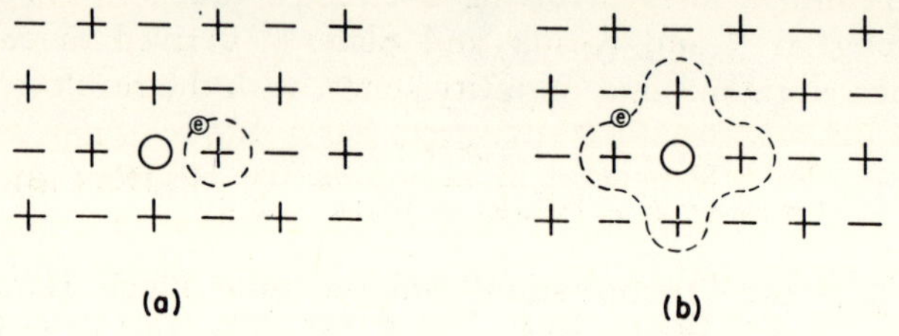

FIG. 14. Schematic picture of (a) early, and (b) more recent versions of the electron-transfer model of the internal structure of the exciton in alkali halide crystals. In the latter version, the electron is shared among alkali ions neighboring the halogen, which is regarded as containing a p^6 shell with a hole in it. This is, in a certain sense, a very-small-radius Wannier exciton.

supported by two observations. First, they were able to predict fairly accurately the energy required in such a process, approximating it by

$$E = E_A - E_I + \alpha_M e^2/a \qquad (5.2)$$

in which E_A is the electron affinity of the halogen (energy required to remove an electron), E_I the ionization energy of the alkali atom (energy gained in returning the electron to the neighbor), α_M the Madelung constant, and a the nearest-neighbor distance. Second, they observed doublet absorption peaks, common to the halide ion, such as the one in KBr (Fig. 1) which, as in the work of Franck and collaborators on alkali halide molecules, could be explained[107] in terms of the existence of "two electron affinities" of the halogen. In other words, the removal of an electron from a filled p^6 shell leaves the p^5 configuration in either a $^2P_{1/2}$ or $^2P_{3/2}$ state, and these states differ in energy by $\lambda = \frac{3}{2}\zeta_p$, where ζ_p is the usual spin-orbit coupling parameter.[108] It is safe to say that this influence of the halogen spin-orbit interaction on alkali halide absorption spectra is their best

[107] See, e.g., J. Franck, H. Kuhn, and G. Rollefson, Z. Physik 43, 155 (1927).
[108] See, e. g., E. U. Condon and G. H. Shortley, "The Theory of Atomic Spectra," p. 122. Cambridge Univ. Press, London and New York, 1953.

understood feature today. The doublet has been observed in nearly all chlorides, and bromides, and iodides.[109,110]

Although Hilsch and Pohl's empirical formula (5.2) does a remarkably good job in predicting the observed absorption energies (see Table III), a formula actually derived from the simple transfer model is[7]

$$E = E_A - E_I + (2\alpha_M - 1) e^2/a \tag{5.3}$$

in which the electrostatic (third) term has a simple origin: an amount of work $\alpha_M e^2/a$ is expended in removing the electron, and another amount $\alpha_M e^2/a$ in bringing it back to the position of a positive ion. However, on the return trip, there is a missing charge whose potential must be subtracted from the Madelung term. As shown in Table III, this more rigorous formula is not as successful as Eq. (5.2). We must regard it as an empirical fact that $2\alpha_M - 1$ has an effective value (not its computed value, because of the simplicity of the model), which happens to be approximately equal to α_M. This effective value is never more than 8 % different from α_M in the examples of Table III.

TABLE III. COMPUTATION OF ELECTRON-TRANSFER EXCITON ENERGIES IN A SELECTED GROUP OF FACE-CENTERED CUBIC ALKALI HALIDES

The first 5 numerical columns are, respectively, the halide electron affinity, the negative of the alkali atom ionization energy, the absolute value of the Madelung energy per ion, and a computation from Wolff and Herzfeld's formula (5.3). The last column is the energy associated with the lowest observed strong absorption peak. All energies are in ev.

	E_A	$-E_I$	$\alpha_M e^2/a$	Hilsch-Pohl	Wolff-Herzfeld	Obs. Lowest level (80° K)
CsF	4.15?	−3.87	8.38	8.7	12.2	9.3
KCl	3.83	−4.32	8.02	7.5	11.0	7.9
NaBr	3.65	−5.12	8.44	7.0	10.6	6.7
RbI	3.14	−4.16	6.88	5.9	8.9	5.7

[109] A study of thin-film absorption spectra of most of the alkali halides at 80° K and 300° K has been done by J. E. Eby, K. J. Teegarden, and D. B. Dutton, *Phys. Rev.* **116**, 1099 (1959); earlier work on the alkali halides was done by E. G. Schneider and H. M. O'Bryan, *Phys. Rev.* **51**, 293 (1937); P. L. Hartman, J. R. Nelson, and J. G. Siegfried, *Phys. Rev.* **105**, 123 (1957); K. Teegarden, *Phys. Rev.* **108**, 660 (1957).

[110] Resolution of the fluoride doublet has thus far been impossible. J. E. Eby *et al.*[109] used LiF as a substrate and did not measure its absorption; this gap has been filled by A. A. Milgram and M. P. Givens, *Phys. Rev.* **125**, 1506 (1962), and R. Onaka, *J. Phys. Soc. Japan* **16**, 340 (1961).

Attempts were made in the 1930's to improve on the simple transfer model,[8,111,112] but agreement with experiment was never really made better than that obtained using Hilsch and Pohl's empirical formula.[113] Klemm[111] included the polarization energy associated with the transfer state, and von Hippel[8] attempted to include the effects of temperature on the transition energy. It is not clear that Klemm's polarization calculations are of use in a modern version of the electron transfer process, because they are based on a model in which electron transfer takes place between two definite ions, creating a real dipole (relative to the ionic lattice). As we will see later, it is probable that the charge distribution in the true excited state has at most a quadrupole moment, in which case Klemm's terms are overestimates of the true polarization energy. Dykman,[114] for example, has assumed that the transfer model implies a multipole consisting of a hole surrounded by six charges of $-e/6$ each.

Mott[113,115] and Seitz[116] pointed out that a state created by transferring an electron from a halogen to a single alkali ion is not a stationary state, even in the absence of translational motion of the excitation. Consider, for example, the NaCl-type lattice or its two-dimensional counterpart in Fig. 14. Transfer of an electron to any of the nearest alkali neighbors of a halogen result in a state of essentially the same energy as any other such state, whereupon the electron finds it easy to jump around from one nearest neighbor to another. To show this quantitatively it would, naturally, be possible to write down a detailed energy matrix for a crystal with a basis, using antisymmetrized product states $\Phi_{\alpha i}(\mathbf{K}, \beta)$, as in Sections 2-4, in which the index i is used to distinguish among localized states a_α or ϕ_α centered on different atoms in the basis. It is perhaps easier first to visualize a Wannier exciton in a diatomic lattice qualitatively: a hole is centered on a point in one sublattice and the electron is described in terms of a hydrogenic envelope of localized orbitals on the other lattice.[117] Assume, then, that the effective exciton radius is very small;

---

[111] W. Klemm, *Z. Physik* **82**, 529 (1933).

[112] M. Born, *Z. Physik* **79**, 62 (1932); J. de Boer, "Electron and Emission Phenomena." Cambridge Univ. Press, London and New York, 1935.

[113] See a discussion in N. F. Mott and R. W. Gurney, "Electronic Processes in Ionic Crystals," pp. 98ff. Oxford Univ. Press, London and New York, 1940:

[114] I. M. Dykman, *Zh. Eksperim. i Teor. Fiz.* **26**, 307 (1954).

[115] See N. F. Mott, *Trans. Faraday Soc.* **34**, 500 (1938).

[116] F. Seitz, "Modern Theory of Solids," p. 411. McGraw-Hill, New York, 1940.

[117] This is an oversimplification of the true situation in a diatomic lattice, made for

so small, in fact, that the usual hydrogenic equation is of little use in determining the lowest state. One possible situation is depicted by Fig. 14b, in which the electron spends its time equally on each nearest-neighbor site but nowhere else. Recall that it was possible to describe the electron in the Wannier exciton by a combination of orbitals (for $\mathbf{K} = 0$) $\psi_\nu = \Sigma_\beta F_\nu(\boldsymbol{\beta}) a_\beta(\mathbf{r})$ [Eq. (4.41)]. Similarly, the small-radius exciton under discussion here can be reasonably described by using a one-electron orbital

$$\psi_1 = (nO)^{-1/2} [a_1(\mathbf{r}) + \cdots + a_n(\mathbf{r})] \tag{5.4}$$

for the electron, where the $a_i(\mathbf{r})$ are localized (Wannier or atomic) functions centered at each of the n nearest neighbors and O is a normalization factor differing from unity because of possible overlapping among the a_i. The "hydrogenic envelope" here is sharply truncated s function.

In the case of an alkali halide, transfer of an electron to a neighbor would be likely to place the electron in the lowest unfilled s orbital of the alkali ion, forming an alkali atom; e.g, in NaCl,

$$a_i(\mathbf{r}) = \psi_{3s}^{Na}(\mathbf{r} - \mathbf{R}_i).$$

If there are six neighbors, there are six possible linear combinations of the a_i having definite transformation properties under the operations of the cubic group. One of these is totally symmetric and is given by Eq. (5.4) with $n = 6$. It behaves like an atomic s function. Three of the others are such that they transform into one another like atomic p functions, and the remaining two transform into one another like the atomic d functions proportional to $3z^2 - r^2$ and $x^2 - y^2$. When these six electron functions are combined with the threefold degenerate p^5 hole function, there result eighteen states of various over-all symmetries, among them appearing two degenerate sets of P-like states. These states are of importance because they may be reached by allowed optical transitions (Section 9). The charge distribution in these states will have a quadrupole moment, not a dipole moment as in the elementary transfer model.

A symmetry analysis of these nearest-neighbor transfer states was

the purpose of this discussion. In practice, it is quite likely that a conduction band Wannier function will have some amplitude on both sublattices. However, it can be defined as being formally localized around the lattice points of one of them.

given by Overhauser[11] in 1956. He showed that the inclusion of both triplet and singlet states increased the number of P-like levels from two, mentioned above, to a total of five. All five of these levels should be reached by allowed optical transitions provided an appreciable spin-orbit coupling is present. Moreover, Overhauser showed that six P-like levels exist in a similar model of the exciton in an ionic lattice of CsCl (simple cubic) type. This increase is due in part to the increase in total electron wave functions available; there are eight nearest neighbors and eight linear combinations like Eq. (5.4) are possible. In particular, there is one [Eq. (5.4) with $n = 8$] of s-like symmetry. In the cesium halides, most of which have the CsCl structure, there do appear to be more absorption peaks clustered about the lowest one than in the case of NaCl type crystals,[6,109,118] and Overhauser concluded on these grounds that the electron-transfer model was in reasonable agreement with experiment. The multiplicity will be discussed again in Sections 5c and 5d.

Various calculations of the exciton excitation energy and line separations on the electron transfer model have been made. Dykman and Tsertsvadze[119] compute an excitation energy of 7.5 ev in NaCl, as compared with the observed[109] peak position at 80° K of about 8.0 ev. Their electron wave function was constructed using a linear combination (5.4) of localized wave functions computed by Tolpygo and Tomasevich.[120] The author is not aware of any further computations using their method, which gives remarkably good agreement with experiment. Very few details of their calculation are available. Goodman and Oen[121] extended Overhauser's formalism by giving the matrix elements of the kinetic and electrostatic energy operators; Overhauser had given only the spin-orbit and overlap matrices. These authors give only a qualitative account of their numerical calculations, in which it is assumed that only nearest neighbor overlaps are non-zero, and the excitation energies were apparently not computed. Pappert[122] developed Overhauser's formalism considerably and

---

[118] F. Fischer and R. Hilsch, *Nachr. Akad. Wiss. Goettingen, II. Math.-Physik. Kl.* No. 8, p. 241 (1959).

[119] I. M. Dykman and A. A. Tsertsvadze, *Zh. Eksperim i Teor. Fiz.* **34**, 1319 (1958); see *Soviet Phys. J.E.T.P. (English Transl.)* **7**, 910 (1958).

[120] K. B. Tolpygo and O. F. Tomasevich, *Ukr. Fiz. Zh.* **3**, 145 (1958); *Fiz. Tverd. Tela* **2**, 3110 (1960); see *Soviet Phys.—Solid State (English Transl.)* **2**, 2765 (1961).

[121] B. Goodman and O. S. Oen, *Phys. Chem. Solids* **8**, 291 (1959).

[122] R. A. Pappert, *Phys. Rev.* **119**, 525 (1960).

applied it to NaCl, taking into account all important overlap integrals, the largest of which were those involving Na 3s functions on nearest like neighbors. His predicted excitation energy of 23.2 ev and the attendant difficulties in computing it led him to the conclusion that unmodified alkali atom functions are generally inappropriate for computations of this type, because of their great spatial extent. Dexter[123] had noticed this earlier, pointing out that the use of unmodified Na 3s functions in NaCl leads to an inconsistency in the transfer model. Whereas the electron is *formally* transferred to the nearest neighbors by combining the 3s functions according to Eq. (5.4), most (72 %) of its charge density lies outside the atomic volumes of the central Cl$^-$ ion and 6 nearest Na$^+$ ions.

c. *The Excitation Model*

Several calculations have been made on the basis of models which we will call collectively "the excitation model." As in the case of the transfer model, attention is paid to setting up appropriate one-electron functions for the electron, which is moving in the potential of the hole, assumed localized at the origin. Taking this potential as $-e^2/\epsilon r$ over most of the crystal, Dexter[10] derived the following electron wave function, valid at large distances from the hole:

$$\psi_\nu(\mathbf{r}) = F_\nu(\mathbf{r}) g(\mathbf{r}) \tag{5.5}$$

where $F_\nu(\mathbf{r})$ is a hydrogenic envelope function modified, as usual, for the dielectric medium, and $g(\mathbf{r})$ is the lowest Bloch state in the conduction band (assumed to lie at $k = 0$). The connection between this function and the results of the Wannier formalism can be obtained quite easily, if we restrict ourselves to $\mathbf{K} = 0$ and use Eq. (4.41) for the orbital centered at the origin:

$$\psi_{\nu 0}(\mathbf{r}) = \sum_\beta{}' F_\nu(\boldsymbol{\beta}) a_\beta(\mathbf{r}). \tag{5.6}$$

Now at large distances $F_\nu(\boldsymbol{\beta})$ can be approximated by $F_\nu(\mathbf{r})$, since the latter will not vary much over the region in which $a_\beta(\mathbf{r})$ is localized. Hence $\psi_{\nu 0}(\mathbf{r})$ becomes approximately

$$F_\nu(\mathbf{r}) \sum_\beta{}' a_\beta(\mathbf{r}). \tag{5.7}$$

---

[123] D. L. Dexter, *Phys. Rev.* **108**, 707 (1957).

Except for a normalization factor, $\Sigma_\beta a_\beta(r)$ is just the Bloch conduction state $g(\mathbf{r})$ for $\mathbf{k} = 0$ [cf. Eq. (2.11)]. Also by analogy with the Wannier model, the excitation model predicts zero-order excitation energies $E_G - G/n^2$ (Section 4). In his calculations on NaCl, Dexter used Tibbs'[124] results for $g(\mathbf{r})$ and a hydrogenic $1s$ envelope function with a damping length of 3 Bohr radii, suggested by Tibbs' detailed F-center calculations. He found oscillator strengths[10] and zero-order excitation energies[123] to be in reasonably good agreement with experiment. The contour of his wave function along the (100) axis is shown in Fig. 15. Since $g(\mathbf{r})$ is a conduction state of a diatomic crystal, $a_\beta(\mathbf{r})$

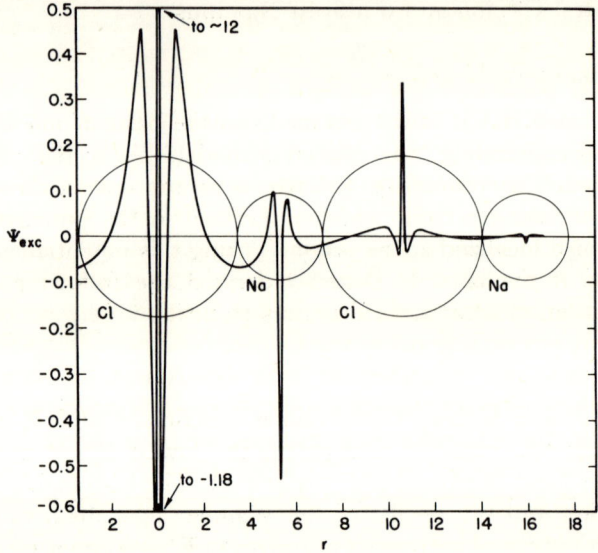

FIG. 15. Contour of the (unnormalized) excitation-model electron wave function in the (100) direction [after D. L. Dexter, *Phys. Rev.* **108**, 707 (1957)]. The Cl⁻ ion at $\mathbf{r} = 0$ is the center of the excitation and the function has cubic symmetry about $\mathbf{r} = 0$.

contains admixtures of atomic-like states from different ions, and $g(\mathbf{r})$ behaves like a Cl⁻⁻ $4s$ function near a Cl nucleus and like a Na⁰ $3s$ function near a Na nucleus. It is thus seen that with respect to localization Dexter's exciton is practically a Frenkel exciton, since 75 % of the electron's charge lies inside the Cl⁻ atomic volume. This

[124] S. R. Tibbs, *Trans. Faraday Soc.* **35**, 1471 (1939).

is somewhat disconcerting, since his electron wave function is known to be correct in detail only at large distance from the hole. It would be valuable to have a computation of the first-order corrections to the excitation energy on this model.

Muto and co-workers[125] computed the binding energy of a hydrogenic $1s$ exciton in all the alkali chlorides, taking into account the details of the Coulomb potential near the origin by using

$$V(\mathbf{r}) = -\frac{1}{\epsilon} \int \frac{|\phi_{3p}(\mathbf{r}')|^2}{|\mathbf{r} - \mathbf{r}'|} d\mathbf{r}' \tag{5.8}$$

and taking the high-frequency dielectric constant for ϵ. Rather than assuming an effective mass for the exciton (or electron, here, since the hole is taken to be fixed), they essentially treated it as a variational parameter (by using a screening factor $Z_{\text{eff}} = \mu/m\epsilon$). In NaCl and KCl, they obtain binding energies of 1.81 and 1.83 ev, corresponding to reduced masses of 1.0 and 0.85 m, respectively. The damping lengths involved are 2.25 a_0 and 2.51 a_0, surprisingly close to the Tibbs-Dexter value of 3 a_0 in NaCl. Muto's early calculations neglect modulation by $g(\mathbf{r})$ entirely, so this cannot be considered equivalent to the excitation energy calculation suggested above; in fact, his electron wave functions are probably not orthogonal to the Cl-core functions. In spite of this, his calculations are in agreement with experiment, if, with Mott,[9] one takes the energy of the second major room-temperature absorption peak in the chlorides as a measure of their direct band gap. This is probably not correct in the light of modern low-temperature data, in which a small "shoulder" at lower energies is thought to give a better indication of the position of the band edge (Section 12). Muto and co-workers have more recently extended their formalism[126,127] to include modulation by the conduction band Wannier functions and have given an absolute estimate of the position of the first exciton level in KCl, which is 8.3 ev (as compared with the experimentally observed 7.9 ev absorption line[109]). This estimate involves assumptions about the electron-hole exchange interaction

---

[125] T. Muto and H. Okuno, *J. Phys. Soc. Japan* **11**, 633 (1956); **12**, 108 (1957); T. Muto and S. Oyama, *ibid.* **12**, 101 (1957).

[126] T. Muto, S. Oyama, and H. Okuno, *Progr. Theoret. Phys. (Kyoto)* **20**, 804 (1958).

[127] T. Muto, *Progr. Theoret. Phys. (Kyoto), Suppl.* **12**, 3 (1959).

which possibly involve several tenths of a volt and which must remain unverified until KCl conduction band Wannier functions are available.

In the preceding discussion of the excitation model, which can be regarded as an $n = 1$ Wannier state, we have perhaps tended to obscure the fact that Wannier excitons higher in the hydrogenic series might be observable in the alkali halides. Fischer and Hilsch[118] do in fact see a second weak absorption peak in KI and RbI at 14° K and assign it to an $n = 2$ state. This interpretation is most interesting and undoubtedly correct, but it is not advisable to deduce exciton binding energies, as these authors do, from only the $n = 1$ and 2 levels. The presence of Wannier excitons in large-band-gap solids appears to have been confirmed in Xe by Baldini,[64] who finds that a definite series exists and that the $n = 1$ state is not accurately positioned in the series. The rare gas spectra are truly a remarkable "intermediate" case since they appear to display both Frenkel and Wannier states! The "Frenkel" state is the "$n = 1$ Wannier" state, which occurs almost precisely at the atomic excitation energy.

d. Multiplicity of Absorption Peaks

We have seen in Section 5b that Overhauser's nearest-neighbor, electron-transfer calculation was able to predict, qualitatively, the number of low-lying absorption lines encountered in crystals of CsCl type as compared with those of NaCl type. This one success is, however, largely overshadowed by the extremely poor results one obtains from numerical applications of the model[123,124] as a result of its spatially diffuse electron function. Moreover, as Dexter[123] and Inchauspé and the author[128] have pointed out, the multiplicity prediction does not really distinguish between the transfer and excitation models. The next higher states on the excitation model will be likely to involve electron symmetries similar to those available from nearest neighbor LCAO's in the transfer model. This fact has been recognized implicitly in calculations by Osaka and colleagues[129] who analyze transfer states by replacing interaction integrals in the Hamiltonian matrix by their atomic p^5s and p^5d counterparts.

---

[128] R. S. Knox and N. Inchauspé, *Phys. Rev.* **116**, 1093 (1959).

[129] Y. Osaka, *J. Phys. Soc. Japan* **14**, 1685 (1959); Y. Osaka, Y. S. Osaka, and F. Goto, *J. Phys. Soc. Japan* **17**, 1715 (1962).

As mentioned earlier, the halogen doublet is probably the best-understood feature of the known alkali halide exciton energy levels. A quantitative study of this doublet cannot help us choose or construct a zero-order model of the exciton, however, because the splitting is determined largely by the hole wave function, which has p character, and which is the common feature of all the zero-order models discussed in Sections 5b and 5c. The doublet is easily understood on any of the preceding models as the result of coupling together the hole and a totally symmetric electron wave function (5.4) or (5.5) with due regard to their spins and spin-orbit coupling of the hole. Two P-like ($J = 1$) states result, whether LS, jj, or intermediate coupling is used. The doublet can even be explained on the Wannier model, inasmuch as we admit that its components correspond to "Wannier 1s excitons," however poorly described quantitatively. The p-like halogen valence band is split into two branches at $\mathbf{k} = 0$ by the spin-

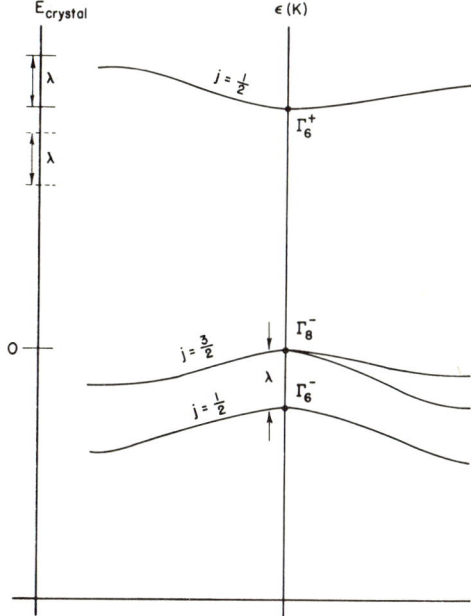

FIG. 16. Possible band structure of an alkali chloride or bromide. At the left are plotted the energy levels of the crystal as a whole; the pair of dashed lines show the position of $\mathbf{K} = 0$ exciton states and the solid lines show the positions of onset of band-to-band transitions.

orbit interaction, which we have ignored in our formalism thus far (see Fig. 16). The "$j = \frac{3}{2}$" (or Γ_8^-) valence band is the higher one, and the lowest-energy exciton state which can be constructed from this band and the conduction band (assumed to be "$j = \frac{1}{2}$" or Γ_6^+-like) gives rise to the lower-energy component of the doublet. We can call this the "$(\frac{3}{2} \frac{1}{2})$" exciton. Clearly, a $(\frac{1}{2} \frac{1}{2})$ exciton can be constructed from the conduction band and the lower valence band. If its binding energy is negligibly different from the first, the upper component of the doublet will lie at an energy λ above the lower component, since a correspondingly larger energy gap must be spanned to create it. It should be noted that we have been speaking of these two excitons as though "pure jj coupling" prevails. Since there can be some exchange interaction $\varDelta$ between the electron and hole, the problem must be treated in intermediate coupling in general, and it is found that the spacing between the doublet components is[128]

$$\delta = 2\left[\tfrac{1}{4}(\varDelta - \tfrac{1}{3}\lambda)^2 + \tfrac{2}{9}\lambda^2\right]^{1/2} \geqslant (\tfrac{8}{9})^{1/2} \lambda; \qquad (5.9)$$

note that δ reduces to λ when the exchange interaction $\varDelta$ is zero. (This expression is not valid when other states of the crystal having over-all P-like symmetry lie close to the doublet.) In the LS coupling

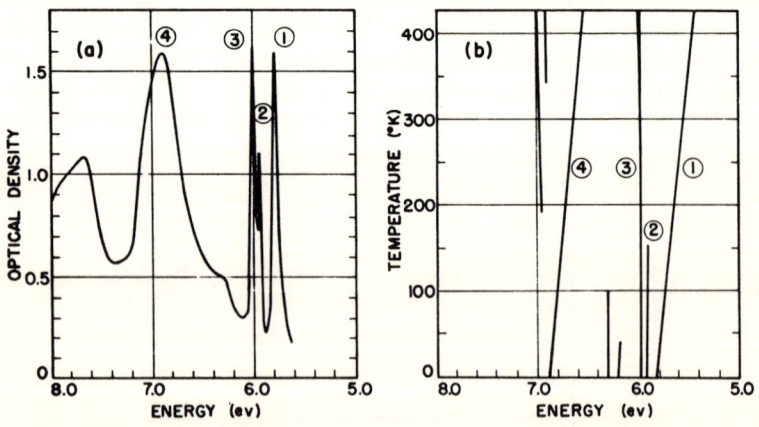

Fig. 17. Unusual temperature dependence of absorption lines in CsI: (a) represents data at 77° K [after J. E. Eby, K. J. Teegarden, and D. B. Dutton, *Phys. Rev.* **116**, 1099 (1959)]; (b) follows F. Fischer and R. Hilsch [*Nachr. Akad. Wiss. Goettingen, II. Math.-Physik. Kl.* No. 8, p. 241, (1959)]; Eby *et al.* also find the structure to be changed at 300° K, in accordance with (b).

limit δ approaches Δ, the energy difference between the purely singlet and triplet excitons. In most of the chlorides and bromides, the inequality (5.9) appears to hold,[109] and δ is close to λ; in the iodides it appears that the valence band splitting is so large that band-to-band transitions originating in the upper valence band set in at lower energies than that of the $(\tfrac{1}{2}\tfrac{1}{2})$ exciton peak.

In concluding this section on multiplicities we would like to call attention to recent work by Fischer and Hilsch.[118,130] On the basis of the temperature dependence of the absorption peaks in CsBr and CsI (see Fig. 17), these authors isolate the halogen doublet and another "doublet" whose spacing corresponds to that of the $^2P_{1/2,\ 3/2}$ doublet in the first excited term of neutral cesium. They give a qualitative explanation of the spectra in terms of LCAO states of the corresponding diatomic molecules, implying that an electron may be transferred from the halide ion to one of the excited states of Cs^0.

e. Prospects for Future Work

The presence of the simple doublet in the chlorides and bromides indicates that a simple electron wave function is involved in the lowest exciton state. Whether it most closely resembles Dexter's damped conduction state, Overhauser's LCAO, Muto's linear combination, or Wannier functions, or something else, it is undoubtedly totally symmetric. This should simplify the calculation of an *in situ* wave function. Although such a calculation is likely to be very informative, it may still be possible to carry on with the electron transfer model by using highly damped and polarized alkali *s* functions,[123] which would lead naturally to the introduction of alkali *p* functions and the possibility of explaining Fischer and Hilsch's observation. At present, the most we dare ask for is a consistent electronic model on which both the excitation energy and oscillator strength of even one alkali halide absorption line can be computed to within reasonable accuracy.

Recently Bassani[131] has computed $\mathbf{k}=0$ band gaps of several alkali halides using the orthogonalized plane wave method in a fairly crude approximation and has obtained good order-of-magnitude agreement with experimental band gaps (i.e., $E_G \sim 1$ ev higher than the lowest exciton absorption energy). The point of greatest interest

[130] F. Fischer, *Z. Physik* **160**, 194 (1960).
[131] F. Bassani, Unpublished data, 1962.

to exciton theory in Bassani's calculation is that the Madelung energy $-\alpha_M e^2/a$ appears only *once* in the expression for the band gap; this arises from the fact that the valence band wave function is taken to be tightly bound, feeling the Madelung energy only at the halide sites, whereas the conduction band wave function is more nearly a plane wave, which samples an average (zero) Madelung energy. Bassani's calculation may provide the first definite progress in understanding the Hilsch-Pohl empirical rule on the basis of the excitation model, since this model leans heavily on knowledge of the band gap for its prediction of excitation energy. It remains to be shown that the exciton binding energy is not affected greatly by first-order Madelung corrections.

6. Effects of Static External Fields

In the preceding three sections we have studied excitons as quantum-mechanical states of a perfect isolated crystal described by a Hamiltonian $\mathcal{H}_0$. Before considering the interactions of these states with photons and phonons, certain simpler perturbations will be examined. These are the static perturbations produced by electric, magnetic, and strain fields. Actually, these fields merely reduce the symmetry of the preceding problem, and their effects are generally small enough to be considered safely as first-order perturbations. The importance of static external fields to exciton theory derives from their ability to probe symmetries of the unperturbed states.

a. *Electric Fields*[53,132]

When the crystal is placed in a uniform external electric field **E** in the z direction, the Hamiltonian $\mathcal{H}_0$ [Eq. (2.2)] becomes

$$\mathcal{H} = \mathcal{H}_0 - e\mathsf{E} \sum_i z_i \qquad (6.1)$$

where z_i is the z component of the position coordinate of the ith electron. If we regard the term in **E** (the "Stark" term) as a perturbation on the states calculated in earlier sections, we need matrix elements of this term between the ground state Φ_0 and the exciton

---

[132] E. F. Gross, B. P. Zakharchenya, and N. M. Reinov, *Dokl. Akad. Nauk S.S.S.R.* **97**, 57, 221 (1954).

states $\Psi_{\nu K}$, and between different exciton states. Since the stark term is odd under inversion of all electron coordinates, it can connect only states of opposite parity, and in particular has zero expectation value in the ground state. Furthermore, since the Stark term is unchanged (except for irrelevant additive factors) during any coordinate translation, it connects only those states of the same translational symmetry. Thus we need consider only those matrix elements of the form

$$\int \Psi_{\lambda K}^* \sum_i z_i \Psi_{\lambda' K} \, d\tau_1 \cdots d\tau_{2N} \tag{6.2}$$

and

$$\int \Psi_{\lambda'' 0}^* \sum_i z_i \Phi_0 \, d\tau_1 \cdots d\tau_{2N} \tag{6.3}$$

where λ and λ' are quantum numbers describing states of opposite parities and λ'' is of odd parity. Finally, since $\Sigma_i z_i$ consists of one-electron operators, the matrix element (6.2) will be zero unless either the electron or the hole is common to excitons λ and λ'. In the notation of Section 2, we obtain for Eq. (6.2)

$$\sum_\beta U_{\lambda K}(\beta)^* z_{mm'}(\beta) U_{\lambda' K}(\beta) \tag{6.4}$$

where

$$z_{mm'}(\beta) = \int a_{m\beta}(\mathbf{r})^* z a_{m'\beta}(\mathbf{r}) \, d\mathbf{r} \tag{6.5}$$

and m and m' label the bands or atomic states used in constructing the exciton states λ and λ'. If the hole is common to λ and λ', then m and m' label electron states; if the electron is common to λ and λ', then m and m' label hole states. In obtaining Eq. (6.4), we have ignored all two-center matrix elements of z. The final Stark perturbation matrix element is

$$- e\mathsf{E} \sum_\beta U_{\lambda K}(\beta)^* z_{mm'}(\beta) U_{\lambda' K}(\beta). \tag{6.6}$$

It will now be estimated for three possible cases of interest, namely the Frenkel exciton, the Wannier exciton with $m \neq m'$, and the Wannier exciton with $m = m'$.

In the Frenkel case, we have $U(\beta) = \delta_{\beta 0}$, and the Stark perturbation is immediately $- e\mathsf{E}z(0)$, where $z(0)$ is obtained from Eq. (6.5) by setting $\beta = 0$ and identifying the a's with atomic functions ϕ.

Thus, if we are concerned with the mixing of $3p^54s$ and $3p^54p$ excitons in solid argon, for example, the matrix element is

$$- e\mathsf{E} \int \phi_{4s}(\mathbf{r})^* z\phi_{4pz}(\mathbf{r})\, d\mathbf{r} \qquad (6.7)$$

which is of order $- e\mathsf{E} \times 5a_0 \sim - 0.3 \times 10^{-7}\, \mathsf{E}$ ev, if E is expressed in volts per centimeter. For normal field strengths, this perturbation is entirely too small to be observable in the rare gases or alkali halides. As is well known from the hydrogen-atom Stark effect, a vanishing energy denominator is needed for linear Stark mixing, and the $4s$-$4p$ energy difference here may be of the order of tenths of an electron volt. By extending the preceding calculation to include second-order perturbations on the ground state, it may be verified quite easily that, in a crystal whose exciton states are best described by the Frenkel or excitation model, the quadratic Stark effect expected in absorption spectra is essentially the same as that for a free atom or molecule. Thus, only small Stark shifts are generally expected to occur in these cases. These shifts will be masked by line broadening in atomic and ionic crystals, where half-widths are of the order of tenths of electron volts, but may be observable in molecular crystals. The author is not aware of any experiments along these lines in molecular crystals.

In the Wannier case with $m \neq m'$ (interband mixing), we denote Eq. (6.5) by $z_{mm'}$, since it is seen to be independent of $\boldsymbol{\beta}$. Equation (6.3) then reduces to

$$z_{mm'} \sum_{\boldsymbol{\beta}} U_{\lambda\mathbf{K}}(\boldsymbol{\beta})^* U_{\lambda'\mathbf{K}}(\boldsymbol{\beta}). \qquad (6.8)$$

This quantity vanishes if $z_{mm'}$ vanishes; if $z_{mm'}$ does not vanish, the sum in Eq. (6.8) may still vanish if the envelope functions U have different symmetry (angular quantum numbers); if (say) they are both $1S$ functions, the sum is appreciable but not precisely equal to 1. The radial parts of the hydrogenic envelope functions are different if the λ and λ' excitons have different effective masses. For all practical purposes, however, Eq. (6.8) may be replaced by $z_{mm'}\, \delta_{\lambda\lambda'}$, where λ and λ' are the quantum numbers of the hydrogenic envelope functions U_λ and $U_{\lambda'}$. The order-of-magnitude estimates for the Frenkel exciton apply here, too; this matrix element will normally be very small in comparison with the energy difference between the two excitons being connected and will lead to small quadratic effects.

It is clear that the search for Stark shifts comparable to absorption line widths must be made in crystals where there already exists a near degeneracy in states of opposite parity. One such case is the 2S-2P degeneracy in the hydrogenic exciton series arising from a particular pair of valence and conduction bands. Let us therefore now look at the Wannier case with $m = m'$, i.e., when the excitons connected by the Stark perturbation are built from the same bands. To study this case, we change the variables of integration in Eq. (6.5) from $\mathbf{r}$ to $\mathbf{r}' = \mathbf{r} - \boldsymbol{\beta}$, obtaining

$$z_{mm}(\boldsymbol{\beta}) = \int a_{m\boldsymbol{\beta}}(\mathbf{r}')^* (z' + \beta_z) a_{m\boldsymbol{\beta}}(\mathbf{r}') \, d\mathbf{r}' \tag{6.9}$$

where β_z is the z component of $\boldsymbol{\beta}$. The contribution from z' is zero whenever the point group of the crystal contains the inversion, which we will assume to be the case. Thus $z(\boldsymbol{\beta}) = \beta_z$ and the Stark matrix element becomes

$$- e\mathbf{E} \sum_{\boldsymbol{\beta}} U_{mn\nu\mathbf{K}}(\boldsymbol{\beta})^* \beta_z U_{mn\nu'\mathbf{K}}(\boldsymbol{\beta}). \tag{6.10}$$

If the hydrogenic model has been used to compute the coefficients U as functions of a quasi-continuous variable $\boldsymbol{\beta}$, we may consistently replace Eq. (6.10) by

$$- e\mathbf{E} \int F_\nu(\mathbf{r})^* z F_{\nu'}(\mathbf{r}) \, d\mathbf{r} \tag{6.11}$$

in which the F are hydrogenic envelope functions with standard normalization. Equation (6.11) is the result we would have obtained by treating the exciton simply as a pair of effective particles. The effective two-particle exciton Hamiltonian in Eq. (4.1) would be augmented by a term

$$- e\mathbf{E} \cdot \mathbf{r}_1 + e\mathbf{E} \cdot \mathbf{r}_2 = - e\mathbf{E} \cdot \mathbf{r} \tag{6.12}$$

where $\mathbf{r}$ is the relative coordinate [Eq. (4.4)].

Apparently the first experimental work on the exciton Stark effect was done by Gross and colleagues,[132] who applied fields of up to about 3×10^4 v/cm to Cu_2O crystals. Their early results qualitatively established the presence of the effect, along with that of field ionization of the exciton.[133] Considerably more quantitative work on the effect was done by Thomas and Hopfield,[89] who studied the nearly degenerate 2S and 2P levels of one of the exciton series in the hexagonal

[133] F. Seitz, *Phys. Rev.* **76**, 1376 (1949).

crystal CdS (see Section 4c). In Fig. 18 we show the main features of their results, compared with the analogous effect in the hydrogen atom. The zero-field energy levels in CdS are already quite different from those in hydrogen; the crystal field does not have inversion

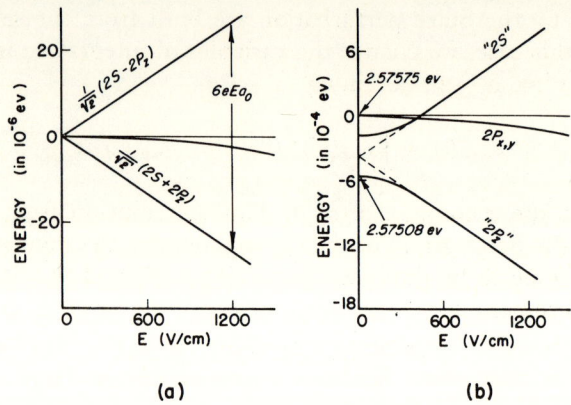

FIG. 18. The Stark effect in the $n = 2$ levels of (a) a hydrogen atom, (b) an exciton in CdS (see text). Part (b) follows D. G. Thomas and J. J. Hopfield [*Phys. Rev. Letters* **5**, 505 (1960); *Phys. Rev.* **124**, 657 (1961)]. The hydrogenic quantum numbers in (b) refer to the envelope symmetry. The z direction coincides with the c axis, and (x, y) lie in the hexagonal plane. The states "$2S$" and $2P_{x,y}$ are not resolved except at fields above ~ 800 v/cm.

symmetry, and $2S$ and $2P_z$ mix with each other and move away from $2P_x$ and $2P_y$. Furthermore, the usual short-range (first-order) perturbations on the Wannier exciton presumably help to lift the $2S$-$2P$ degeneracy.

As an electric field in the z direction is applied to a hydrogen atom,[134] the $2s$ and $2pz$ states mix and are repelled linearly with a separation $6e\mathbf{E}a_0$, as sketched in Fig. 18a. The $2px$ and $2py$ states suffer only a small quadratic shift. These effects can be seen quite clearly in CdS (Fig. 18b). Stark shifts at small fields will generally be quadratic in the case of the nondegeneracy of $2s$ and $2pz$; a quasi-linear portion will appear as soon as the Stark perturbation is of the same order of magnitude as the zero-field splitting. Note the change of scale between Figs. 18a and b. The effective Bohr radius of the

[134] See, e.g., L. I. Schiff, "Quantum Mechanics," 2nd ed., pp. 158-160. McGraw-Hill, New York, 1955.

exciton in CdS is 29 times that of hydrogen, so at 1000 v/cm the quantity $6e\,\mathsf{E}a_0$ is 10^{-3} ev, readily observable in Thomas and Hopfield's experimental arrangement. It should be emphasized that a Stark measurement such as this one will not generally be possible on an arbitrarily chosen crystal. In CdS, we are fortunate to have $n = 2$ exciton states whose symmetry properties are such that weak, sharp lines appear in a relatively uncluttered region of the optical absorption spectrum when the polarization vector of the exciting light is parallel to the c axis. The CdS $2S$ and $2P_z$ states discussed above are, in fact, "longitudinal" excitons. The Stark effect in the corresponding $1S$ exciton state is too small to be observed, and the effect in higher states of the series is washed out by field ionization.

Field ionization, well known in atomic physics, occurs when $e\,\mathsf{E}d$, the potential energy difference across the diameter d of an orbit, is of the order of magnitude of the binding energy of a particle in the orbit. At fields large enough for this to be the case, the lifetime of the state against dissociation becomes quite short, and the line width for absorption to that state becomes correspondingly large. In CdS, Thomas and Hopfield find that excitons of higher radius ($n \geqslant 3$) broaden and merge into the continuum at relatively low fields (~ 500 v/cm). Gross and Zakharchenya[135] find that higher fields (~ 9000 v/cm) are necessary to do the same in the yellow series of Cu_2O. The Bohr radius of this exciton in Cu_2O is $\sim 17\,a_0$, or about one-half that of the CdS exciton under consideration, and its binding energy is some four times as large. Thus, field about eight times as large should be required to destroy the $n = 3$ exciton. The experimental results at least show qualitative agreement with this rough estimate. Allen[136] has considered the ionizing effect of the electric fields associated with grain boundaries (Section 12).

b. Magnetic Fields[53,89,137–139]

Magnetic perturbations on the crystal involve one-electron operators, and their formal calculation using general many-electron exciton

---

[135] E. F. Gross and B. P. Zakharchenya, *Zh. Tekhn. Fiz.* **28**, 231 (1958); see *Soviet Phys.—Tech. Phys. (English Transl.)* **3**, 206 (1958).

[136] J. W. Allen, *Nature* **187**, 51 (1960).

[137] L. L. Korenblit, *Zh. Tekhn. Fiz.* **27**, 484 1957); see *Soviet Phys.—Tech. Phys.* **2**, 434 (1957).

[138] R. J. Elliott and R. Loudon, *Phys. Chem. Solids* **8**, 382 (1959).

[139] R. J. Elliott and R. Loudon, *Phys. Chem. Solids* **15**, 196 (1960).

states would parallel the first part of Section 6a closely. We thus consider immediately the two-particle model of the Wannier exciton, which suffices to explain magnetic effects wherever they have thus far been observed. In the presence of a magnetic field derived from a vector potential **A**, the Schrödinger equation (4.1) for the electron and hole becomes[140]

$$\left[-\frac{\hbar^2 \nabla_e^2}{2m_e^*} - \frac{\hbar^2 \nabla_h^2}{2m_h^*} - \frac{e^2}{\epsilon r_{eh}} + \frac{ie\hbar}{m_e^* c} \mathbf{A}(\mathbf{r}_e) \cdot \nabla_e \right.$$
$$\left. - \frac{ie\hbar}{m_h^* c} \mathbf{A}(\mathbf{r}_h) \cdot \nabla_h + \frac{e^2}{2m_e^* c^2} \mathbf{A}(\mathbf{r}_e)^2 + \frac{e^2}{2m_h^* c^2} \mathbf{A}(\mathbf{r}_h)^2 \right] \Psi = \mathsf{E}\Psi. \quad (6.13)$$

We will consider only uniform magnetic fields, in which case the vector potential is taken to be

$$\mathbf{A}(\mathbf{r}) = \tfrac{1}{2}(\mathbf{H} \times \mathbf{r}). \quad (6.14)$$

If we now make a transformation to center-of-mass and relative coordinates,

$$\boldsymbol{\rho} = (m_e^* \mathbf{r}_e + m_h^* \mathbf{r}_h)/(m_e^* + m_h^*), \quad (6.15)$$

$$\mathbf{r} = \mathbf{r}_e - \mathbf{r}_h \quad (6.16)$$

and make a canonical transformation[141] equivalent to assuming that Ψ has the form

$$\Psi(\boldsymbol{\rho}, \mathbf{r}) = \exp\left\{i[\mathbf{K} - (e/\hbar c)\mathbf{A}(\mathbf{r})] \cdot \boldsymbol{\rho}\right\} F(\mathbf{r}) \quad (6.17)$$

we obtain the following effective Schrödinger equation for F:

$$\left[-\frac{\hbar^2}{2\mu}\nabla^2 - \frac{e^2}{\epsilon r} + \frac{ie\hbar}{c}\left(\frac{1}{m_e^*} - \frac{1}{m_h^*}\right)\mathbf{A}(\mathbf{r}) \cdot \nabla \right.$$
$$\left. + \frac{e^2}{2\mu c^2}\mathbf{A}(\mathbf{r})^2 - \frac{2e\hbar}{(m_e^* + m_h^*)c}\mathbf{A}(\mathbf{r}) \cdot \mathbf{K}\right]F = \left[E - \frac{\hbar^2 K^2}{2(m_e^* + m_h^*)}\right]F.$$
$$(6.18)$$

As in Section 4, $1/\mu = 1/m_e^* + 1/m_h^*$. In addition to the unperturbed

---
[140] The convention used throughout this article is such that e is an algebraic number, i.e., the charge on the electron is $e = -4.8 \times 10^{-10}$ esu.
[141] W. Lamb, *Phys. Rev.* **85**, 259 (1952).

6. EFFECTS OF STATIC EXTERNAL FIELDS

effective Hamiltonian appearing in Eq. (4.10), three magnetic terms appear. The first is the usual Zeeman term

$$\frac{ie\hbar}{c}\left(\frac{1}{m_e^*} - \frac{1}{m_h^*}\right) \mathbf{A}(\mathbf{r}) \cdot \nabla = -\frac{e}{2c}\left(\frac{1}{m_e^*} - \frac{1}{m_h^*}\right)\mathbf{H} \cdot \mathbf{L} \quad (6.19)$$

where $\mathbf{L} = \mathbf{r} \times (-i\hbar\nabla)$ is the relative angular momentum operator. The second is the diamagnetic term

$$\frac{e^2 \mathbf{H}^2(x^2 + y^2)}{8\mu c^2} \quad (6.20)$$

here written for $\mathbf{H}$ in the z direction. The third is a term arising from the quasi-electric field which an observer riding with the center of mass of the exciton would experience because of the magnetic field in the laboratory[142]:

$$-\frac{2e\hbar}{(m_e^* + m_h^*)c}\mathbf{A}(\mathbf{r}) \cdot \mathbf{K} = -\left[\frac{\hbar \mathbf{K}}{(m_e^* + m_h^*)} \times \mathbf{H}\right] \cdot \mathbf{r} = -\frac{1}{c}(\mathbf{v} \times \mathbf{H}) \cdot \mathbf{r} \quad (6.21)$$

where $\mathbf{v} \equiv \hbar\mathbf{K}/(m_e^* + m_h^*)$ is the velocity of the exciton. Although the present calculation presupposes a parabolic dependence of the exciton energy on $\mathbf{K}$, a more general treatment would presumably show that the velocity appearing in Eq. (6.21) is to be taken as $\hbar^{-1}\nabla_\mathbf{K} E(\mathbf{K})$.

The use of these three terms as perturbations on hydrogenic states, the theory of which is well known from atomic spectroscopy, requires that the magnetic terms be small compared with the Coulomb term $-e^2/\epsilon r_{eh}$. When the latter is in fact very small compared with the magnetic terms,[143] this perturbation procedure is not valid, and, better, $-e^2/\epsilon r_{eh}$ might be considered as the perturbation. Obviously the unperturbed solutions are then products of Landau states for the electron and hole. Elliott and Loudon[138,139] have computed eigenfunctions and eigenvalues for the high magnetic field case; their results are of considerable interest to the theory of interband magnetoabsorption, and will be discussed in Section 9.

---

[142] This term appears to have been noted first for excitons by Korenblit, [137] who underestimated its size by several orders of magnitude.

[143] This condition has obtained in the case of impurity states in semiconductors; see Y. Yafet, R. W. Keyes, and E. N. Adams, *Phys. Chem. Solids* **1**, 137 (1956).

Exciton levels in the presence of magnetic fields have been studied in Cu_2O,[144,145] Ge,[146-149] CdS,[53,89,150,151] and CdSe.[95,152,153] Generally, it can be concluded that the Wannier model is confirmed in detail in these crystals, but several highlights of the experiments and interpretations deserve mention. In Cu_2O, Gross and colleagues find that the diamagnetic term (6.20) contributes an observable shift in the $n = 3$ level at fields of about 30 kilogauss. At $n = 5$, the order of magnitude of this shift agrees with values predicted from exciton radii known from the optical spectrum. In hydrogen, because of the much smaller electron orbits, this effect can be observed only in very high members of the series ($n \gtrsim 30$). While demonstrating the presence of a diamagnetic effect, Gross also noted the absence of a linear Zeeman effect which might have been expected to result from the $\mathbf{H} \cdot \mathbf{L}$ term (6.19). Gross took this as an indication that m_e^* and m_h^* were approximately equal, in which case Eq. (6.16) is very small. Hopfield and Thomas[49] pointed out that the linear Zeeman effect in exciton P states will be quenched by the longitudinal-transverse exciton splitting if the magnetic field is perpendicular to the exciton wave vector, but their numerical estimates failed to account for Gross's observation in Cu_2O. It should be noted that Eq. (6.19) accounts only for the *orbital* contribution to the Zeeman splitting. In all but states which are pure spin singlets, the interaction of the spins of the

---

[144] E. F. Gross and B. P. Zakharchenya, *Zh. Tekhn. Fiz.* **27**, 1940 (1957); see *Soviet Phys.—Tech. Phys. (English Transl.)* **2**, 1802 (1957); E. F. Gross, B. P. Zakharchenya, and P. P. Pavinskii, *Zh. Tekhn. Fiz.* **27**, 2177 (1957); see *Soviet Phys.—Techn. Phys. (English Transl.)* **2**, 2018 (1957); E. F. Gross, B. P. Zakharchenya, and A. I. Sibilev, *Fiz. Tverd. Tela* **4**, 1003 (1962) [*English Transl: Sov. Phys. Solid State* **4**, 739 (1962)].

[145] E. F. Gross, *Phys. Chem. Solids* **8**, 172 (1959).

[146] S. Zwerdling, L. M. Roth, and B. Lax, *Phys. Rev.* **109**, 2207 (1958).

[147] S. Zwerdling, B. Lax, L. M. Roth, and K. Button, *Phys. Rev.* **114**, 80 (1959); S. Zwerdling, L. M. Roth, and B. Lax, *Phys. Chem. Solids* **8**, 397 (1959).

[148] K. Button, L. M. Roth, W. H. Kleiner, S. Zwerdling, and B. Lax, *Phys. Rev. Letters* **2**, 161 (1959).

[149] D. F. Edwards and V. J. Lazazzera, *Phys. Rev.* **120**, 420 (1960).

[150] R. G. Wheeler, *Phys. Rev. Letters* **2**, 463 (1959); R. G. Wheeler and J. O. Dimmock, *Phys. Rev. Letters* **3**, 372 (1959).

[151] E. F. Gross, B. P. Zakharchenya, and B. S. Razbirin, *Fiz. Tverd. Tela* **3**, 3083 (1961) [*English Transl.: Sov. Phys. Solid State* **3**, 2243 (1962)].

[152] J. O. Dimmock and R. G. Wheeler, *J. Appl. Phys.* **32**, 2271 (1961).

[153] J. J. Hopfield, *J. Appl. Phys.* **32**, 2277 (1961).

6. EFFECTS OF STATIC EXTERNAL FIELDS

electron and hole with the magnetic field must be considered. This results in the addition of a term

$$g_e \mu_B \cdot \mathbf{S}_e + g_h \mu_B \cdot \mathbf{S}_h \tag{6.22}$$

to the Hamiltonian (6.13). Here $\mu_B = e\mathbf{H}/mc$, and the electron and hole spins and g values are denoted by $\mathbf{S}_e$, $\mathbf{S}_h$, g_e, and g_h. The effective g value for any given exciton state will be of the form

$$g_{\text{eff}} = a_\mu g_\mu + a_e g_e + a_h g_h \tag{6.23}$$

where the a's are numerical coefficients which depend on the symmetries of the hole, electron, and envelope function, and

$$g_\mu \equiv m \left(\frac{1}{m_h^*} - \frac{1}{m_e^*} \right). \tag{6.24}$$

In the presence of appreciable spin-orbit coupling g_μ may not be given accurately by the effective masses in the valence and conduction bands, but may at least be regarded as an adjustable parameter. Hopfield and Thomas have given a list of these effective g values in CdS.[53] For the $2P(\Gamma_{15}^-)$ exciton in Cu_2O, assuming that the yellow series arises from a Γ_7^+ valence band and a Γ_6^+ conduction band, we obtain

$$g_{\text{eff}} = \tfrac{1}{4} | 2g_\mu - 2g_e + \tfrac{2}{3} g_h |. \tag{6.25}$$

It is likely that the effective g is the one which actually vanishes in Cu_2O, so that m_e^* and m_h^* are not necessarily equal. Figure 19 shows a typical Zeeman effect in CdS.

Two types of excitons are known to exist in germanium, as remarked in Section 4c. Experiments using large magnetic fields (of the order of 40 kilogauss) at Lincoln Laboratory showed that the $\mathbf{K} = 0$ level (at 0.882 ev in zero field) shifted quadratically with field strength, indicating a diamagnetic exciton state. Confirmation of the interpretation of band edge data in terms of a $\mathbf{K} = (111)$ exciton by Macfarlane et al.[97] was also provided by this work. Since the binding energy of the $\mathbf{K} = 0$ exciton is very small, a hydrogenic series is not resolved and the position of the band gap must be obtained indirectly. The Lincoln Laboratory group used extrapolated Landau level data, obtaining $E_G(\mathbf{k} = 0) = 0.898$ ev at low temperatures and 0.889 ev at 77° K. Macfarlane et al.[97] and Edwards and Lazazzera[149] have, on the other hand, analyzed the zero-field absorption curves according to Elliott's formalism (see Section 9) and find $E_G(\mathbf{k} = 0) = 0.883$ ev at 77° K. Edwards and Lazazzera's data, which is claimed to be taken on unstrained samples, displays Landau levels whose energies, as a

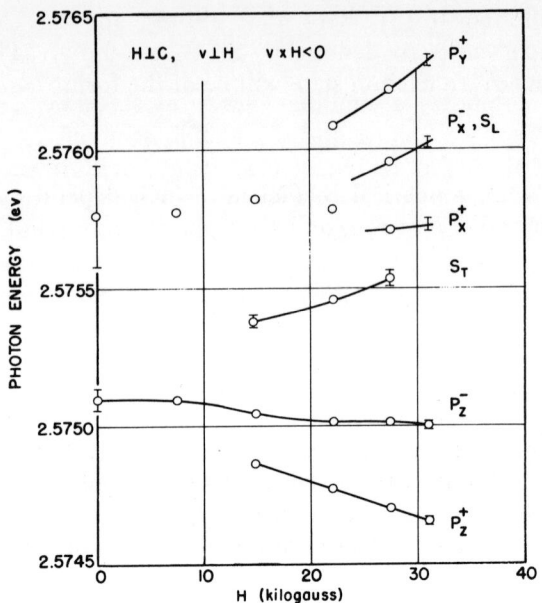

FIG. 19. Zeeman effect on the $n = 2$ excitons of the "A" series in a thin crystal of CdS [after D. G. Thomas and J. J. Hopfield, *Phys. Rev. Letters* **5**, 505 (1960); *Phys. Rev.* **124**, 657 (1961)]. S and P refer to hydrogenic envelopes, as in Fig. 18b. L and T refer to longitudinal and transverse states, as discussed in the text. Superscripts denote spin states split by the magnetic field. The separation between P_z^+ and P_y^+ at 31 kilogauss is shown in Fig. 20 as a function of an additional electric field.

function of magnetic field, appear to extrapolate linearly to a point *lower* than the zero-field exciton energy. These authors point out that, if these "Landau states" were to contain an admixture of exciton states, their energies might contain a quadratic term and pull up above the true exciton line at zero field. They consider their results a confirmation of Elliott and Loudon's calculations, which predict that the traditional Landau levels as observed in semiconductors not only contain an admixture of exciton states, but are *mainly* exciton-like. The true Landau state is to appear in absorption as a small line on the high-energy tail of the main absorption.

Similar considerations apply to the indirect exciton, but thus far the Landau extrapolation is the only independent means of obtaining the corresponding gap energy. The Lincoln group finds E_G (indirect, Ge) = 0.744 ev at low temperatures.

The possibility of observing a third type of exciton in Ge has been considered by Lax.[154] This one would have $\mathbf{K} = 0$ but would originate from electrons and holes with $\mathbf{k} = (111)$, thus, a "direct (111) transition." Lax computes a binding energy of 0.005 ev, and suggests that this exciton will show up in reflection at about 2 ev, the direct band gap at the point L. Direct (111)-edge excitons have also been mentioned in connection with absorption edge data in other semiconductors.[155]

The exhaustive analysis of CdS by Hopfield and Thomas is remarkable on several counts. CdS is an anisotropic crystal in which effective mass and g tensors have more than one component, and these authors have been able to assign a reasonably unique set of values to all their components, explaining a variety of quantitative experiments. Furthermore, interesting magnetic effects were observed, such as the longitudinal-transverse quenching of the Zeeman effect and the interchanging of intensities of absorption lines on reversal of a magnetic field; the latter will be considered in Section 9. Finally, they were able to detect the effect of the Lorentz term (6.21) on the CdS spectrum. A constant electric field $\mathbf{E}$ affects the two-particle Hamiltonian (6.13) by adding a term (6.12) which combines with Eq. (6.21) to produce a total perturbation

$$-\left(\mathbf{E} + \frac{1}{c}\mathbf{v}\times\mathbf{H}\right)\cdot\mathbf{r}. \qquad (6.26)$$

Using the $2P_z - 2P_y$ energy difference (Fig. 18b) as a measure of the effective Stark field, Thomas and Hopfield turned on a magnetic field of 31 kilogauss and then essentially balanced out $(\mathbf{v}/c)\times\mathbf{H}$ by varying $\mathbf{E}$. As seen in Fig. 20, a field of about 100 v/cm is required, which implies $v/c \sim 10^{-5}$. The effective mass $m_e^* + m_h^*$ computed from this velocity is 1.1 electron masses, which agrees quite well with the sum of the electron and hole effective masses appropriate to the exciton under study (0.20 and 0.7, respectively). Thomas and

---

[154] B. Lax, *Phys. Rev. Letters* **4**, 511 (1960).
[155] M. Cardona and G. Harbeke, *Phys. Rev. Letters* **8**, 90 (1962).

Hopfield have thus used a magnetic field to probe the *translational* symmetry of the exciton as well as its internal structure.

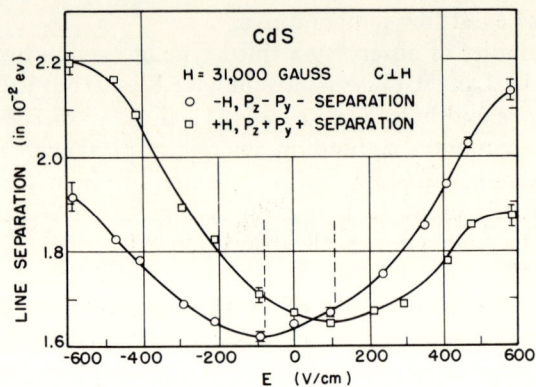

FIG. 20. Magnetostark effect in CdS [after D. G. Thomas and J. J. Hopfield, *Phys. Rev. Letters* **5**, 505 (1960); *Phys. Rev.* **124**, 657 (1961)]. The curve marked by □'s is the 31-kilogauss $P_z^+ - P_y^+$ separation noted in Fig. 19, now as a function of E. At E ∼ 100 v/cm, this separation is reduced to the magnitude it would presumably have in an exciton with no velocity and no applied electric field. The curve marked by ○'s is taken with **v** × **H** > 0, in which case a different but essentially equivalent pair of magnetic levels are observed in optical absorption and used for the measurement.

c. Strain Fields

Strains have generally been of secondary importance to exciton theory in the sense that their primary influence is felt on the one-electron bands, whose masses and degeneracies are changed under compression or shear. These effects are reflected in the structure of exciton levels in the strained crystals, making additional experimental parameters available to check symmetry and mass assignments in the one-electron bands. An exception to this is the work of Gross and co-workers[82,156,157] on the $n = 1$ and 2 levels of the Cu_2O

---

[156] E. F. Gross and A. A. Kaplianskii, *Fiz. Tverd. Tela* **2**, 1676, 2968 (1960) [*English Transl.: Sov. Phys. Solid State* **2**, 1518, 2637 (1961)]; E. F. Gross, B. P. Zakharchenya, and L. M. Kanskaya, *Fiz. Tverd. Tela* **3**, 972 (1961) [*Englsh Transl.: Sov. Phys. Solid State* **3**, 706 (1961)].

[157] E. F. Gross, A. A. Kaplianskii, and V. T. Agekyan, *Fiz. Tverd. Tela* **4**, 1009 (1962) [*English Transl.: Sov. Phys. Solid State* **4**, 744 (1962)]; E. F. Gross, A. A. Kaplianskii, V. T. Agekyan, and D. S. Bulyanitsa, *Fiz. Tverd. Tela* **4**, 1660 (1962) [*English Transl.: Sov. Phys. Solid State* **4**, 1219 (1962)].

yellow series; here, deformations oriented along specific crystallographic directions have been extremely helpful in elucidating the symmetry of the $1S$ exciton (quadrupolar), which in turn has made serious speculation on the Cu_2O band structure possible.[70]

Deformation potentials in germanium were derived by Kleiner and Roth[158] from data on direct exciton level shifts in strained samples. These authors assumed as a perturbation on the Wannier Hamiltonian the sum of the one-electron valence and conduction band strain Hamiltonians. Thomas[159] has assigned a symmetry to the valence band of CdTe by inference from strained exciton measurements. Similarly, strain effects on CdSe have been studied.[160]

7. Special Topics

Although the exciton might well have been the by-product of modern many-body theory, it grew up, in its own right, roughly as outlined in Sections 3-5. The purpose of this section is to indicate some connections which exist between conventional exciton theory and modern treatments of many-body problems, without going into a detailed mathematical discussion of the latter.[161] We will be particularly interested in the effective electron-hole interaction.

a. Second Quantization and Exciton Statistics[162]

For many purposes it is convenient to systematize the handling of determinantal wave functions by using the techniques of the theory of quantized fields. Our vacuum state Φ_0 will be defined as a single determinant describing a perfect nonmetal with a filled valence band, Eq. (2.5), and the operator $c_{n\mathbf{k}}$ destroys an electron (creates a hole) in band n with wave vector $\mathbf{k}$. Similarly, $c_{n\mathbf{k}}^\dagger$ creates an electron in this

---

[158] W. H. Kleiner and L. M. Roth, *Phys. Rev. Letters* **2**, 334 (1959).
[159] D. G. Thomas, *J. Appl. Phys.* **32**, 2298 (1961).
[160] L. T. Chadderton, R. B. Parsons, W. Wardzynski, and A. D. Yoffe, *Phys. Chem. Solids* **23**, 416 (1962).
[161] For recent reviews of the methods of many-body theory, see D. J. Thouless, "Quantum Mechanics of Many-Body Systems." Academic Press, New York, 1961.
[162] This section is necessarily abbreviated. See L. I. Schiff, "Quantum Mechanics," 2nd ed. McGraw-Hill, New York, 1955.

state. Thus, as an example, an alternative form for the state $\Psi_{mn}(\mathbf{k}_e, \mathbf{k}_h)$ [see Eq. (2.17)] is

$$c^\dagger_{n\mathbf{k}_e} c_{m\mathbf{k}_h} \Phi_0. \tag{7.1}$$

Regarding the electronic structure of the crystal as a field whose zero-order, ground-state wave function is Φ_0, we compute its first-order energy as the expectation value of the operator

$$H = \int d\mathbf{r}\, \Psi(\mathbf{r})^\dagger H_1(\mathbf{r}) \Psi(\mathbf{r}) + \tfrac{1}{2} \iint d\mathbf{r}\, d\mathbf{r}'\, \Psi(\mathbf{r})^\dagger \Psi(\mathbf{r}')^\dagger \frac{e^2}{|\mathbf{r} - \mathbf{r}'|} \Psi(\mathbf{r}') \Psi(\mathbf{r}) \tag{7.2}$$

in which

$$H_1(\mathbf{r}) = \frac{p^2}{2m} + V(\mathbf{r}) \tag{7.3}$$

is the effective one-electron Hamiltonian for the Bloch states, and the Ψ are operators expanded in any complete set of functions, which for convenience may be taken as the one-electron eigenfunctions of H_1:

$$\Psi(\mathbf{r}) = \sum_{n\mathbf{k}} \psi_{n\mathbf{k}}(\mathbf{r})\, c_{n\mathbf{k}} \tag{7.4}$$

$$\Psi(\mathbf{r})^\dagger = \sum_{n\mathbf{k}} \psi_{n\mathbf{k}}(\mathbf{r})^*\, c^\dagger_{n\mathbf{k}}. \tag{7.5}$$

Computation of $(\Phi_0, H\Phi_0)$ with due regard for the anticommutation relations

$$c_{n\mathbf{k}} c^\dagger_{n'\mathbf{k}'} + c^\dagger_{n'\mathbf{k}'} c_{n\mathbf{k}} = \delta_{nn'} \delta_{\mathbf{k}\mathbf{k}'}$$

$$c_{n\mathbf{k}} c_{n'\mathbf{k}'} + c_{n'\mathbf{k}'} c_{n\mathbf{k}} = 0$$

$$c^\dagger_{n\mathbf{k}} c^\dagger_{n'\mathbf{k}'} + c^\dagger_{n'\mathbf{k}'} c^\dagger_{n\mathbf{k}} = 0 \tag{7.6}$$

results in the expression (2.8) for $E_0^{(1)}$, with $U = V$.

We may similarly define creation and annihilation operators for the localized one-electron states, i.e., $c^\dagger_{n\mathbf{R}}$ and $c_{n\mathbf{R}}$; in terms of these operators, the state $\Psi_{mn}(\mathbf{R}_i, \mathbf{R}_j)$ [Eq. (2.22)] becomes

$$c^\dagger_{n\mathbf{R}_j} c_{m\mathbf{R}_i} \Phi_0. \tag{7.7}$$

The Wannier-state operators $c^\dagger_{n\mathbf{R}}$ and $c_{n'\mathbf{R}'}$ obey anticommutation relations analogous to Eq. (7.6), as may be readily verified from the equations connecting the two sets [cf. Eq. (2.14)]:

$$c^{\dagger}_{n\mathbf{R}} = N^{-1/2} \sum_{\mathbf{k}} e^{-i\mathbf{k}\cdot\mathbf{R}} c^{\dagger}_{n\mathbf{k}}. \tag{7.8}$$

The calculations of Section 2 were concerned with finding linear combinations of pair states (7.1) or (7.7), constructed using two specific bands m and n, which would diagonalize H. The resulting states were shown to be mixtures of states of the same total wave vector $\mathbf{K} = \mathbf{k}_e - \mathbf{k}_h$, for which it was convenient to define an exciton representation [Eqs. (2.27), (2.28)]. Here it is then convenient to define an analogous exciton creation operator

$$b^{\dagger}_{\boldsymbol{\beta}\mathbf{K}} = N^{-1/2} \sum_{\mathbf{k}} e^{-i\boldsymbol{\beta}\cdot\mathbf{k}} c^{\dagger}_{n\mathbf{k}} c_{m,\mathbf{k}-\mathbf{K}} \tag{7.9}$$

$$= N^{-1/2} \sum_{\mathbf{R}} e^{i\mathbf{K}\cdot\mathbf{R}} c^{\dagger}_{n,\mathbf{R}+\boldsymbol{\beta}} c_{m\mathbf{R}} \tag{7.10}$$

and its Hermitian conjugate,

$$b_{\boldsymbol{\beta}\mathbf{K}} = N^{-1/2} \sum_{\mathbf{R}} e^{-i\mathbf{K}\cdot\mathbf{R}} c^{\dagger}_{m\mathbf{R}} c_{n\mathbf{R}+\boldsymbol{\beta}}. \tag{7.11}$$

The operator (7.9) or (7.10) creates an electron-hole pair separated by the vector $\boldsymbol{\beta}$, in a state of total momentum $\mathbf{K}$. We have suppressed the indices (mn) in $b^{\dagger}$ for simplicity. The general exciton (2.31) is thus created by an operator

$$b^{\dagger}_{\nu\mathbf{K}} = \sum_{\boldsymbol{\beta}} U_{\nu\mathbf{K}}(\boldsymbol{\beta}) b^{\dagger}_{\boldsymbol{\beta}\mathbf{K}}. \tag{7.12}$$

It can be seen quite easily that excitons are *almost* bosons. When one attempts to show that $b^{\dagger}_{\nu\mathbf{K}}$ and $b_{\nu'\mathbf{K}'}$ commute, he finds

$$[b_{\nu\mathbf{K}}, b^{\dagger}_{\nu'\mathbf{K}'}] = \delta_{\nu\nu'} \delta_{\mathbf{K}\mathbf{K}'} + O\left(\frac{\text{no. of electron-hole pairs}}{N}\right). \tag{7.13}$$

The second term on the right side of Eq. (7.13) is a result of interference between the excitons, and would not occur if it were possible to "doubly excite" a lattice site. The determinantal form of the wave function of a crystal containing two or more Frenkel excitons must be normalized carefully to account for this interference.[40,163,164]

Provided that the number of excitons remains small compared with N, the number operator

$$N_{\nu \mathbf{K}} = b^\dagger_{\nu \mathbf{K}} b_{\nu \mathbf{K}} \qquad (7.14)$$

has eigenvalues 0, 1, 2, ···, corresponding to the number of excitons of type ν present with wave vector $\mathbf{K}$. Similarly, the total energy of the crystal relative to the ground state (vacuum) is

$$\sum_{\nu \mathbf{K}} b^\dagger_{\nu \mathbf{K}} b_{\nu \mathbf{K}} E_{\nu \mathbf{K}}, \qquad (7.15)$$

where $E_{\nu \mathbf{K}}$ is the energy of a single $\nu \mathbf{K}$-type exciton.

The use of the notation of second quantization is convenient in discussing interactions of excitons with particles and fields and we shall use it in future sections. Although operator techniques could be employed to re-derive the first-order results of Section 2, it is not clear that much would be gained thereby, since the major problem (in practice, at the moment) has been found to be the actual determination of good one-electron wave functions, whether localized states in tight-binding calculations, or band structures in the case of the Wannier exciton. Explicit consideration of the diagrams of importance to conventional exciton theory has been given by Izuyama[165] and Nozieres and Balkanski.[166] Goodman and Oen[121] have sketched a method by which nonorthogonal basis functions can be used in the second-quantized exciton formalism.

The boson-like nature of excitons has been the basis for speculations on how the exciton's statistics might manifest themselves in phase changes at low temperatures. Simple Bose-Einstein condensation would occur for excitons at and below the critical temperature[167] $T_0 = (1.8\, \hbar^2/M^* k)\, n^{2/3}$, where $M^* = m_e^* + m_h^*$, k is Boltzmann's

---

[163] Y. Toyozawa, *Progr. Theoret. Phys. (Kyoto)* **12**, 421 (1954).

[164] The case of a crystal containing atoms with energy levels so spaced that two excitations can "co-exist" there has been studied by R. E. Merrifield, *J. Chem. Phys.* **31**, 522 (1959).

[165] T. Izuyama, *Progr. Theoret. Phys. (Kyoto)* **22**, 681 (1959).

[166] P. Nozieres and M. Balkanski, Contract AF61(052)-130, Tech. Notes 3 and 4.

[167] See e.g., C. Kittel, "Elementary Statistical Physics," p. 100. Wiley, New York, 1958.

constant, and n is the density of excitons; this density must be $n = 4 \times 10^{17}$ cm^{-3} to produce a critical temperature of 4.2° K for excitons of $M^* = 2m_e$. A density of this order would not necessarily imply that excitons were not expected to behave as bosons [cf. Eq. (7.13)]. To determine whether such densities are attainable one must know the lifetime of the exciton, which is a strong function of its momentum and energy (Sections 10, 12); in particular, the lifetime must be fairly long ($\sim 10^{-6}$ sec). Apparently exciton condensation was first considered by Moskalenko and other Russian workers,[168] who have developed the theory by analogy with that of liquid helium, and who propose to detect the condensation through certain "opticohydrodynamic" phenomena such as "thermal superconductivity." Other authors[169] are fairly optimistic about obtaining high exciton densities, particularly in view of high photon fluxes currently attainable with lasers. They propose to detect the condensation by a series of phenomena based on the assumption that the state into which the excitons condense is weakly coupled to the lattice. In Section 9 we mention a possible complication introduced by strong exciton-photon coupling.

b. The Effective Electron-Hole Interaction

As discussed in Section 4, it is quite likely that at large separations r the effective interaction between an excess electron and a free hole in a solid is $-e^2/\epsilon r$, where ϵ is the low-frequency dielectric constant of the solid. Although no rigorous derivation of this interaction has apparently been given, strong support for the conjecture is provided from three independent theoretical approaches. The first is most closely related to traditional exciton theory and can even be introduced in terms of the two-band model, then readily generalized. We noted in Sections 2 and 4 that in the context of a perturbation-theoretic treatment, higher excited states of the crystal must be used to properly describe the polarization induced by the electron and hole. There are higher zero-order states even within the two-band model,

---

[168] S. A. Moskalenko, *Fiz. Tverd. Tela* **4**, 276 (1962); see *Soviet Phys.—Solid State (English Transl.)* **4**, 199 (1962); and *Opt. Spektroskopiya* **5**, 147 (1958); S. V. Vonsovskii and M. S. Svirskii, *Zh. Eksperim. i Teor. Fiz.* **36**, 1259 (1959); see *Soviet Phys. J.E.T.P. (English Transl.)* **9**, 894 (1959).

[169] J. M. Blatt, K. N. Boer, and W. Brandt, *Phys. Rev.* **126**, 1691 (1962); R. C. Casella, *Phys. Chem. Solids* **24**, 19 (1963).

namely those in which two or more excited pairs are present. Let us consider with Toyozawa[163] the "electronic polaron," which results from the interaction of an electron with an electronically polarizable lattice. The Hamiltonian of the crystal plus an excess charge q (with canonical coordinates $\mathbf{p}, \mathbf{x}$) is given by

$$\mathcal{H}_0 + \frac{p^2}{2m} + U(\mathbf{x}) \tag{7.16}$$

where

$$U(\mathbf{x}) = -\sum_i \frac{q|e|}{|\mathbf{x} - \mathbf{r}_i|} + \sum_I \frac{z_I q|e|}{|\mathbf{x} - \mathbf{R}_I|} \tag{7.17}$$

is the Coulomb interaction between the extra charge and the particles comprising the crystal. An important property of $U(\mathbf{x})$ is that it connects only those states of the crystal whose total exciton populations differ by one. In particular, the matrix element between the ground state and a state containing a single Frenkel exciton is computed to be

$$\int \Phi_0^* \, U(\mathbf{x}) \, \Psi_{l\mathbf{K}} \, d\tau_1 \cdots d\tau_N = q \, \frac{1}{\sqrt{N}} \sum_{\mathbf{R}} e^{i\mathbf{K}\cdot\mathbf{R}} \left[\frac{\boldsymbol{\mu}_l \cdot (\mathbf{x} - \mathbf{R})}{|\mathbf{x} - \mathbf{R}|^3} \right] \equiv e^{i\mathbf{K}\cdot\mathbf{x}} \, \Gamma_{l\mathbf{K}} \tag{7.18}$$

where $\boldsymbol{\mu}_l$ is the (assumed nonvanishing) dipole matrix element connecting the ground and lth excited state of an atom, having the value

$$\boldsymbol{\mu}_l = \int \phi_l(\mathbf{r})^* \, e\mathbf{r} \, \phi_0(\mathbf{r}) \, d\mathbf{r}. \tag{7.19}$$

in the adequate approximation of one electron per atom. Now, regarding the Frenkel excitons as a boson field, the Hamiltonian of the whole system (excess charge plus bosons) is

$$\frac{p^2}{2m} + U_0(\mathbf{x}) + \sum_{l\mathbf{K}} b_{l\mathbf{K}}^\dagger b_{l\mathbf{K}} E_{l\mathbf{K}} + \sum_{l\mathbf{K}} (\Gamma_{l\mathbf{K}} b_{l\mathbf{K}} e^{i\mathbf{K}\cdot\mathbf{x}} + \Gamma_{l\mathbf{K}}^* b_{l\mathbf{K}}^\dagger e^{-i\mathbf{K}\cdot\mathbf{x}}), \tag{7.20}$$

where $U_0(\mathbf{x})$ is the Coulomb interaction of q with the ground-state charge density of the crystal, and the third term is the Hamiltonian operator for the crystal in the absence of the extra charge [i.e., Eq. (7.15)].

For small K, $\Gamma_{l\mathbf{K}}$ may be computed in the continuum approximation (Section 3; Heller and Marcus[39]), and it is found that only longitudinal excitons interact with q, so the second sum in Eq. (7.20) may be accordingly restricted. This fact is a consequence of the longitudinal nature of the Coulomb interaction and has its analog in polaron theory, where only LO (longitudinal optical) phonons interact with an extra electron. Its value in this case is

$$\Gamma_{l\mathbf{K}} = - \frac{4\pi q n_0 \mu_l}{\sqrt{N}} \cdot \frac{i}{K} = - q\gamma \frac{i}{K} \qquad (7.21)$$

where n_0 is the density of lattice points, $\gamma = 4\pi\mu_l \Omega_0^{-1/2}$, and Ω_0 is the volume of a unit cell. Toyozawa showed that γ could be related to the dielectric constant ϵ_∞ by a variational technique used by Frohlich et al.[170] for the polaron; the result was

$$\gamma^2 = 2\pi E_{l\mathbf{K}} \left(1 - \frac{1}{\epsilon_\infty}\right), \qquad (7.22)$$

where $E_{l\mathbf{K}}$ is the energy of a longitudinal Frenkel exciton. Toyozawa "clothed" an electron with a cloud of excitons, using the method of Lee et al.[171] He was primarily interested in the interaction of the Frenkel exciton with an impurity center, and we will return to this work in Section 12. For the present, we are interested in the application of his method by Haken and Schottky[75] to the Wannier exciton. These authors assumed that an excess electron and a hole introduced into a crystal interact with each other according to $- e^2/|\mathbf{x}_e - \mathbf{x}_h|$ and with the crystal according to the fourth term of the Hamiltonian (7.20). Since $\Gamma_{l\mathbf{K}}$ (hole) $= - \Gamma_{l\mathbf{K}}$ (electron), the total Hamiltonian is

$$\frac{p_e^2}{2m} + \frac{p_h^2}{2m} + U_0(\mathbf{x}_e) + U_0(\mathbf{x}_h) - \frac{e^2}{|\mathbf{x}_e - \mathbf{x}_h|} + \sum_{l\mathbf{K}} b_{l\mathbf{K}}^\dagger b_{l\mathbf{K}} E_{l\mathbf{K}}$$

$$+ \sum_{l\mathbf{K}} [\Gamma_{l\mathbf{K}} b_{l\mathbf{K}} (e^{i\mathbf{K}\cdot\mathbf{x}_e} - e^{i\mathbf{K}\cdot\mathbf{x}_h}) + \text{h.c.}]. \qquad (7.23)$$

The first five terms are seen to be the Wannier exciton Hamiltonian in the simple two-particle approximation, if it is recalled that the effect of U_0 on each particle is to change m to an effective mass. Haken and Schottky eliminated the Frenkel exciton operators from

---

[170] H. Frohlich, H. Pelzer, and S. Zienau, Phil. Mag. [7] 41, 221 (1950).
[171] T. D. Lee, F. E. Low, and D. Pines, Phys. Rev. 90, 297 (1953).

the sixth and seventh terms in Eq. (7.23) using products of Lee-Low-Pines states, reducing them to an additional effective potential

$$\left(1 - \frac{1}{\epsilon_\infty}\right) \frac{e^2}{|\mathbf{x}_e - \mathbf{x}_h|} [1 - \tfrac{1}{2}(e^{-\rho_e|\mathbf{x}_e - \mathbf{x}_h|} + e^{-\rho_h|\mathbf{x}_e - \mathbf{x}_h|})] \qquad (7.24)$$

where $\rho_i = (2m_i E_{l\mathbf{K}}/\hbar^2)^{1/2}$. The details of this calculation are given by Haken[172] in earlier papers, dealing with electron-phonon interactions, but using the same technique. Obviously, at large electron-hole separations the potential (7.24) merely introduces ϵ_∞ into the denominator of the electron-hole interaction in Eq. (7.23), while at small separations it is roughly constant and equal to

$$\tfrac{1}{2}\left(1 - \frac{1}{\epsilon_\infty}\right) e^2(\rho_e + \rho_h). \qquad (7.25)$$

The dielectric constant then disappears again from the Coulomb law. Haken and Schottky's derivation is based on the approximation that the momentum of the clothed electron and hole states used in constructing the exciton function are nearly zero. This approximation is expected[172] to break down when $|\mathbf{x}_e - \mathbf{x}_h| \gtrsim \rho_i^{-1}$, but clearly it is not extremely bad at small values of $|\mathbf{x}_e - \mathbf{x}_h|$, since the constant (7.25) turns out to be just the negative of the sum of the self-energies of the clothed electron and hole. At small electron-hole separations the polarization clouds have canceled each other out and the self-energy of the system is eliminated, as we would have hoped.

As Haken and Schottky fully realized, it is inconsistent to assume the existence of a given set of excitons, or "pair state waves," in their terminology, as members of the complete set of solutions of the problem, in order to compute an effective Hamiltonian which will lead to crystal excitations of a different type. The philosophy of their approach is to characterize the polarizability of the lattice by a set of oscillators, represented by the boson field; the bosons to which the electron and hole are in fact assumed to be linearly coupled are not likely to be localized pair states, but might be other excitations, say the Wannier excitons themselves. All that enters quantitatively is the coupling constant, which is determined empirically in this simple model from the dielectric constant itself. Thus exactly the same formalism has been used[172] to extend the formulation of the dielectric constant

[172] H. Haken, Z. Naturforsch. 9a, 228 (1954); Nuovo Cimento [10] 3, 1230 (1956).

problem to the case of lattice polarizability. This amounts to assuming in Eq. (7.23), that the "bare" Coulomb interaction is $-e^2/\epsilon_\infty |\mathbf{x}_e - \mathbf{x}_h|$, and that the bosons of importance in the interaction (seventh) term are longitudinal optical phonons of frequency $\omega_{LO}(\mathbf{K})$ rather than longitudinal excitons. The theory is quantitatively identical to the one just described, with the exception that the following replacements are made everywhere:

$$E_{l\mathbf{K}} \to \hbar\omega_{LO}(\mathbf{K}),$$

$$\left(1 - \frac{1}{\epsilon_\infty}\right) \to \left(\frac{1}{\epsilon_\infty} - \frac{1}{\epsilon_0}\right).$$

Thus, e.g., the effective Coulomb interaction becomes $-e^2/\epsilon_0 |\mathbf{x}_e - \mathbf{x}_h|$ when $|\mathbf{x}_e - \mathbf{x}_h|$ is greater than the larger of the $\rho_i = \sqrt{2m_i\hbar\omega_{LO}/\hbar^2}$. There does not seem to have been any effort to combine these two treatments to obtain a full picture of the effective interaction, probably because little new would be learned. An excellent summary of Haken's work and its relationship to earlier theories is found in his 1958 review.[26]

The second theory relevant to the effective dielectric constant is that of Kohn,[173] who showed that two charges at large distances in a crystal interact according to $q_1 q_2 e^2/\epsilon r_{12}$, where ϵ is the low-frequency dielectric constant. His theory does not apply directly to excitons because at least one of the charges involved must be a classical point charge. Kohn's proof uses the complete set of Bloch states in a linked-cluster perturbation calculation,[174] and is valid whenever certain terms in the perturbation expansion proportional to powers of q_1 and q_2 higher than the first are negligible. These terms contribute short-range interactions, but an explicit radius at which they become nonnegligible is not given. Kohn's calculation is valid for an exciton in which the hole is definitely trapped and therefore is in a sense distinguishable from the other holes appearing in the excitations of the system. In the otherwise perfect crystal, this situation obtains only when $m_h^* \to \infty$. A calculation using somewhat similar techniques, but apparently not yet general enough to include the exciton problem, is described by Roth and Pratt.[175]

---

[173] W. Kohn, *Phys. Rev.* **110**, 857 (1958); an attempt to extend Kohn's considerations to polar crystals has been made by S. J. Nettel, *Phys. Rev.* **128**, 2573 (1962).

[174] K. A. Brueckner, *Phys. Rev.* **100**, 36 (1955); J. Goldstone, *Proc. Roy. Soc.* **A293**, 267 (1957).

[175] L. M. Roth and G. W. Pratt, Jr., *Phys. Chem. Solids* **8**, 47 (1959).

The third relevant theoretical structure is that of Englert,[76] who has studied the problem of small quantized systems residing in a weakly dissipative medium. He computes an effective Hamiltonian for an exciton state of zero wave vector constructed from states (7.1) with $\mathbf{k}_e = \mathbf{k}_h \equiv \mathbf{k}$, the Bloch states concerned being associated with simple spherical bands. The result is that the states $c_{n\mathbf{k}}^\dagger c_{m\mathbf{k}} \Phi_0$ of various $\mathbf{k}$ are coupled by

$$H(\mathbf{k}) = \frac{\hbar^2 k^2}{2\mu} - \sum_q \frac{4\pi e^2}{Vq^2} \frac{1}{\epsilon(q,0)} \exp(\mathbf{q} \cdot \nabla_\mathbf{k}) \qquad (7.26)$$

where V is the volume of the system, Σ_q runs over all wave vectors, and $\epsilon(\mathbf{q}, 0)$ is the zero-frequency, zero-temperature dielectric function[176]

$$\epsilon(\mathbf{q}, 0) = 1 - \sum_s \frac{4\pi e^2}{Vq^2} |\langle s | \rho_\mathbf{q} | 0 \rangle|^2 \left(\frac{1}{E_{s0} + i\eta} + \frac{1}{E_{s0} - i\eta} \right) \qquad (7.27)$$

where Σ_s runs over all true eigenstates of the crystal, $\rho_\mathbf{q}$ is the density operator $\rho_\mathbf{q} = \Sigma_j e^{-i\mathbf{q}\cdot\mathbf{x}_j}$, E_{s0} is the energy difference between the state s and the ground state, and η is a parameter which is eventually allowed to go to zero. Transforming Eq. (7.26) to an exciton representation (7.9) with $\mathbf{K} = 0$, we find that the effective Hamiltonian is

$$-\frac{\hbar^2}{2\mu} \nabla_\beta^2 - \sum_q \frac{4\pi e^2}{Vq^2} \frac{1}{\epsilon(\mathbf{q},0)} e^{-i\mathbf{q}\cdot\boldsymbol{\beta}} \qquad (7.28)$$

which may be written

$$-\frac{\hbar^2}{2\mu} \nabla_\beta^2 - \sum_q \frac{1}{\epsilon(\mathbf{q},0)} \left\{ \frac{e^2}{\beta} \right\}_q \qquad (7.29)$$

where $\{e^2/\beta\}_q$ is the qth Fourier component of e^2/β. If $\epsilon(\mathbf{q}, 0)$ were constant, Eq. (7.29) would be the familiar two-particle Hamiltonian appearing in (4.10) or (4.28). Englert notes that for large β the only important contributions to the sum on $\mathbf{q}$ are from terms with small $\mathbf{q}$ [cf. Eq. (7.25)] and the resulting interaction is $-e^2/\epsilon(0,0)\beta$. At small β, all regions of $\mathbf{q}$ space contribute equally to the integral. Over most of this region q is large and $\epsilon(\mathbf{q}) \approx 1$, whence the effective interaction is $-e^2/\beta$. Englert's exciton, his "small quantized system," is distin-

---

[176] P. Nozieres and D. Pines, *Nuovo Cimento* [10] **9**, 470 (1958).

guishable from the rest of the crystal in that the exciton-crystal exchange interaction is neglected, although the particles in both systems are of the same kind (electrons).

The three theories just discussed have a common feature; to some extent the electron and hole are distinguishable from the particles in the system which they polarize. This is not simply a matter of neglect of exchange energy, but rather it amounts to an artificial construction of excited states of the crystal. One would like the exciton and its appropriately modified hydrogenic spectrum to appear naturally as an excitation of the whole many-body system. However, Kohn's general charge renormalization conjecture does not appear to have been explicitly verified for excitons. This conjecture states[170] that "Any pair of extra charges, provided they are sufficiently far apart, and whether carried by electrons, holes, foreign particles (e.g., protons), or classical charge points, interact with each other via the same Coulomb interaction as in a vacuum, except that each charge Q must be replaced by an effective charge

$$Q_{\text{eff}} = Q/\epsilon^{1/2}$$

where ϵ is the static dielectric constant."

c. Relationship to Collective Excitations

In normal metals, infinitesimally small energies are required to excite the electronic system. There exist, however, collective excitations[177] known as plasma modes (or plasmons), occurring at energies of the order of 10 to 20 ev. These states are the solid state analog of the oscillations of a free-electron plasma which occur at a frequency $\omega_p = (4\pi n e^2/m)^{1/2}$, which we write

$$1 = \frac{4\pi n e^2}{m} \frac{1}{\omega_p^2}. \qquad (7.30)$$

Here n is the electron density in coordinate space. Equation (7.30) is a special case of the plasma dispersion relation for metals,[174]

$$1 = \frac{4\pi n e^2}{m} \sum_{n',\mathbf{k}} \frac{f_{nn'}(\mathbf{k})}{\omega^2 - \omega_{nn'}(\mathbf{k})} \qquad (7.31)$$

---

[177] For a review of early work on plasma oscillations see D. Pines, *Solid State Phys.* **1**, 367 (1955); see also S. Raimes, *Rept. Progr. Phys.* **20**, 1 (1957).

where $f_{nn'}(\mathbf{k})$ is the oscillator strength of a transition from an occupied one-electron state $\psi_{n\mathbf{k}}$ to a state $\psi_{n'\mathbf{k}}$ at an energy higher by $\hbar\omega_{nn'}(\mathbf{k})$. For free electrons, there are no "higher bands"; the sum in Eq. (7.31) reduces to $\omega^{-2} \Sigma_{n'\mathbf{k}} f_{nn'}(\mathbf{k}) = \omega^{-2}$, because of the one-electron sum rule, and (7.30) is obtained. The solutions of Eq. (7.31) for ω are plasmon frequencies for the metal. These excited states are not populated thermally because of their very high energies, and are generally observed only through their excitation by high-energy particles as seen in energy-loss spectra. However, as virtual states they are extremely important in determining the interactions between charges in a metal (including those between the electrons themselves). They screen the Coulomb interaction almost completely, and thereby account for much of the success that the one-electron approximation has enjoyed in metals. On the face of it, this approximation should be poor, because of the supposedly large perturbations produced by the long-range Coulomb interaction. Thus, although one might conceive of ordinary excitons in metals (states involving holes from an inner shell bound to conduction electrons, for example), such states of excitation would be highly unstable[178,179] because conduction electrons can rush in and screen the electron-hole interaction very rapidly, leading to dissociation of the pairs in a time of the order of ω_P^{-1}, or $\sim 10^{-16}$ sec. It should be noted, however, that Friedel[180,181] has suggested that some features of metallic absorption spectra appear to be closely related to those of the (free) constituent atoms. Friedel contends that the screening (of, say, a hole in a d-like metallic band) is almost fully accomplished by one electron, which goes into an orbital closely resembling the excited atomic orbital. If such a state were to have an appreciable lifetime we could regard it as an exciton.

In insulators and semiconductors, plasmons corresponding to the filled valence band also exist, at an energy given roughly by [177,182,183]

$$\hbar\omega = \hbar\omega_P[1 + (E_G/\hbar\omega_P)^2]^{1/2} \tag{7.32}$$

[178] N. F. Mott, *Proc. Phys. Soc. (London)* **A62**, 416 (1949).

[179] P. Nozieres and D. Pines, *Phys. Rev.* **109**, 1062 (1958).

[180] J. Friedel, *Proc. Phys. Soc. (London)* **65**, 769 (1952); *Phil. Mag.* [7] **43**, 153 (1952).

[181] The author is indebted to Professor J. Korringa for bringing his attention to Friedel's work.

[182] C. Horie, *Progr. Theoret. Phys. (Kyoto)* **21**, 113 (1959).

[183] This relation is valid only when $E_G \ll \hbar\omega_P$ and when the bandwidths of the valence and conduction bands are small. Generally, plasmon energies comprise one of a set of excitation energies to be computed from a dispersion relation, and if $E_G \approx \hbar\omega_P$ the plasma modes become mixed with ordinary one-electron excitations.

where E_G is the band gap and ω_P is to be calculated using the density of valence electrons. At the same time, the few electrons present thermally in the conduction band possess a very low-frequency plasma mode at an ω_P corresponding to their spatial density. In order to describe excitons, which we know can exist in these solids, we must explicitly introduce excitations of pairs, which can be avoided in conventional plasmon theory. There might appear, then, to be a fundamental difference between the two types of excitation. Horie,[182] Guseva and Taluts,[184] and more recently Nozieres and Balkanski,[166] have explored their connection. We mention some of the highlights of Horie's work here. He essentially expands the wave function of the excited solid as a linear combination of pair creation and destruction operators acting on the exact ground-state Ψ_0,

$$\Psi_{\nu \mathbf{k}} = \sum_{\mathbf{k}} \sum_{m,n}{}' (u_{mn\nu\mathbf{K}}(\mathbf{k})\, c^\dagger_{n\mathbf{k}}\, c_{m\mathbf{k}-\mathbf{K}} + v_{mn\nu\mathbf{K}}(\mathbf{k})\, c^\dagger_{m\mathbf{k}}\, c_{n\mathbf{k}-\mathbf{K}})\, \Psi_0, \qquad (7.33)$$

where u and v are coupling coefficients to be determined by solving the general eigenvalue problem for the crystal. By the prime is meant that the sums over m and n are to be restricted to valence and conduction bands, respectively; thus the first term creates pairs and the second destroys them. By making a number of simplifying assumptions, some similar to those made by Takeuti (see Section 5a), Horie obtains a single dispersion relation from which he is able to derive both exciton-like and plasmon-like eigenstates. The former is a state in which it happens that $v \ll u$, and a state resembling the exciton state $b^\dagger_{\nu\mathbf{K}}\Phi_0$ is obtained. That it is an exciton state in the usual sense is inferred from the fact that the Fourier transforms of the u's in Eq. (7.33) obey an equation similar to the Slater-Takeuti equation (5.1), and its eigenvalues are below the gap energy. In the plasmon-like solution, $u \approx v$ and there result states

$$\Psi_{\nu\mathbf{k}} \approx \sum_{\mathbf{k}} \sum_{m,n} (u_{mn\nu\mathbf{K}}(\mathbf{k})\, c^\dagger_{n\mathbf{k}}\, c_{m,\mathbf{k}-\mathbf{K}} + u_{mn\nu\mathbf{K}}(\mathbf{k})^*\, c^\dagger_{m\mathbf{k}}\, c_{n\mathbf{k}-\mathbf{K}})\, \Psi_0, \qquad (7.34)$$

which have the plasmon form and whose eigenvalues are derived from appropriate dispersion relations. The states (7.34) are plasmons because they represent "stationary waves of charge fluctuations."[179] Horie attempts to make a distinction among cases in which either both of these types of solution appear, or only one of them. His

---

[184] G. I. Guseva and G. G. Taluts, *Fiz. Metal. i Metalloved.* **7**, 658 (1959).

approximations are fairly severe and a lot of quantitative work is still necessary before this portion of his results can be utilized.

As the band gap in solids is reduced, it becomes harder for excitons to have appreciable lifetimes. In the first place, the polarizability and dielectric constant ϵ increase with decreasing gap, and the exciton rydberg vanishes as ϵ becomes large. Thus thermalization of the exciton to continuum states becomes very easy. Izuyama[165] uses a one-dimensional model with standard bands to show that the exciton rydberg is always smaller than the energy gap, but this problem deserves further investigation in three-dimensional crystals. The polarizability may in some cases depend primarily on vertical-transition energy denominators, all of which could be large in the presence of a small *indirect* gap. A new type of phase change might take place at low temperatures if the exciton rydberg were to become larger than the gap, i.e., a state containing a finite number of *real* excitons might drop out as the ground state. The plasma effect is the most important temperature-dependent factor limiting its occurrence, however. A small gap means relatively large hole and electron populations, which would reinforce the dielectric screening of the electron-hole interaction.

States which have properties strikingly similar to the traditional exciton are known in superconductors.[25,185] Bardasis and Schrieffer[25] have treated these states in detail, but this work is beyond the scope of the present article. Bardeen-Cooper-Schrieffer quasi-particles[186] are coupled in pairs whose center of mass move with wave vectors **K**, and which have the familiar longitudinal and transverse character. Under certain conditions it is expected that these coupled pairs will represent observable states lying within the superconducting energy gap, while under other conditions, the new states resemble plasmons.

d. *X-ray Excitons*

States of very high excitation energy can be prepared when X-rays

[185] Their existence was apparently first suggested by P. W. Anderson, *Phys. Rev.* **112**, 1900 (1958), and N. N. Bogoliubov, V. V. Tolmachev, and D. V. Shirkov, "A New Method in the Theory of Superconductivity." Consultants Bureau, New York, 1959.

[186] J. Bardeen, L. N. Cooper, and J. R. Schrieffer, *Phys. Rev.* **108**, 1175 (1957); see also J. Bardeen and J. R. Schrieffer, *in* "Progress in Low Temperature Physics" (C. J. Gorter, ed.), Vol. 3, p. 170, esp. pp. 258-261. Wiley (Interscience), New York, 1961.

7. SPECIAL TOPICS

are incident on a crystal. Among them are states arising from a hole in one of the deeper bands of the crystal, along with an electron in the conduction band. Because of the electron-hole Coulomb interaction, these states are undoubtedly exciton-like, and the exciton binding energy must be taken into consideration when experimental absorption or emission curves are interpreted along the lines of the band theory of solids. The complications which these "X-ray excitons" introduce into the X-ray spectroscopy of solids has been discussed at length by Parratt.[187]

X-ray excitons must have binding energies of much the same order of magnitude as ordinary excitons; because the hole effective mass is expected to be very large, the exciton reduced mass will be roughly equal to the electron effective mass. This is borne out in Muto's variational calculation for the alkali halides,[126] which he has applied to X-ray as well as ordinary excitons. Other, mainly experimental, papers relevant to this relatively unexplored field of exciton physics are by Mott and Cauchois,[188] Kiyono,[189] Parratt and Jossem,[190] and Johansson.[191]

---

[187] L. G. Parratt, *Rev. Mod. Phys.* **31**, 616 (1959).
[188] N. F. Mott and Y. Cauchois, *Phil. Mag. First Ser.* **40**, 1260 (1950).
[189] S. Kiyono, *Sci. Rept. Tohoku Univ.*, **36**, 1 (1952).
[190] L. G. Parrattt and E. L. Jossem, *Phys. Rev.* **97**, 916 (1955); *Phys. Chem. Solids* **2**, 67 (1957).
[191] P. Johansson, *Arkiv Fysik* **18**, 289 (1960).

III. Absorption and Dispersion of Light by Nonmetallic Solids

8. Classical Theory of the Optical Effects

Because most observations which have been made on excitons have involved optical absorption and related phenomena, we include below for convenience a brief review of the relevant macroscopic theory. Furthermore, since the exciton represents one of the "oscillators" envisaged by Lorentz in his microscopic theory of the optical effects, this theory is surveyed in Section 8b in order to provide an additional link, if only a qualitative one, between exciton theory and the macroscopic theory. The connection is ultimately provided by the well-known characterization of quantum mechanical excited states of atoms by oscillators of charge $ef^{1/2}$, where f is the "strength" of the oscillator.

a. Phenomenological Theory

In an uncharged polarizable medium of negligible magnetic susceptibility, the electric and magnetic field strengths obey Maxwell's equations in the form

$$\nabla \times \mathbf{H} - \frac{1}{c}\frac{\partial(\boldsymbol{\epsilon}\cdot\mathbf{E})}{\partial t} = \frac{4\pi}{c}\boldsymbol{\sigma}\cdot\mathbf{E}$$

$$\nabla \times \mathbf{E} + \frac{1}{c}\frac{\partial \mathbf{H}}{\partial t} = 0$$

$$\nabla \cdot (\boldsymbol{\epsilon}\cdot\mathbf{E}) = 0$$

$$\nabla \cdot \mathbf{H} = 0 \tag{8.1}$$

where $\boldsymbol{\epsilon}$ and $\boldsymbol{\sigma}$ are the dielectric and conductivity tensors, respectively. For simplicity we consider only the steady-state case of a plane monochromatic electromagnetic wave originating in vacuum and incident on a semi-infinite isotropic or cubic dispersive and absorbing medium. Thus, in either medium we take

$$\mathbf{E} = \mathbf{E}_0 \, e^{-i(\boldsymbol{\kappa}\cdot\mathbf{r}-\omega t)} \tag{8.2a}$$

and

$$\mathbf{H} = \mathbf{H}_0 \, e^{-i(\boldsymbol{\kappa}\cdot\mathbf{r}-\omega t)} \tag{8.2b}$$

where $\mathbf{E}_0$ and $\mathbf{H}_0$ are constant complex amplitudes and $\boldsymbol{\kappa}$, the wave vector in the region under consideration, may be complex, in which case its imaginary part provides attenuation of the wave. Substituting Eqs. (8.2) into (8.1) we obtain

$$\frac{c\boldsymbol{\kappa}}{\omega} \times \mathbf{H}_0 = -\left(\epsilon + \frac{4\pi\sigma}{i\omega}\right)\mathbf{E}_0 \tag{8.3a}$$

$$\frac{c\boldsymbol{\kappa}}{\omega} \times \mathbf{E}_0 = \mathbf{H}_0 \tag{8.3b}$$

and

$$\boldsymbol{\kappa}\cdot\mathbf{E}_0 = \boldsymbol{\kappa}\cdot\mathbf{H}_0 = 0. \tag{8.3c}$$

It is conventional to define a complex index of refraction $N = c\kappa/\omega = n - ik$, whose real part n is the ordinary index and in which k is known as the extinction coefficient. Since, according to Eqs. (8.3), the three vectors $\boldsymbol{\kappa}$, $\mathbf{E}_0$, and $\mathbf{H}_0$ are mutually perpendicular, we let $\boldsymbol{\kappa}$ lie along the x axis, $\mathbf{E}_0$ along y, and $\mathbf{H}_0$ along z. Then Eqs. (8.3) lead to $N\mathbf{E}_{0y} = \mathbf{H}_{0z}$ and $N\mathbf{H}_{0z} = [\epsilon + (4\pi\sigma/i\omega)]\,\mathbf{E}_{0y}$, from which we conclude that N must satisfy

$$N^2 = \epsilon + \frac{4\pi\sigma}{i\omega}. \tag{8.4}$$

The two real equations corresponding to Eq. (8.4) are

$$n^2 - k^2 = \epsilon$$

and

$$nk = 2\pi\sigma/\omega. \tag{8.5}$$

When σ is small, N is real and $n \approx \epsilon^{1/2}$. N^2 can be used formally as a complex dielectric constant provided that σ is not introduced as an additional independent parameter.

In applications of exciton theory we often wish to compute the absorption coefficient of a medium, defined in the present case by the

8. CLASSICAL THEORY OF THE OPTICAL EFFECTS

relative decrease of energy flow per unit distance in the direction of propagation, i.e., by

$$\frac{d}{dx}\langle S_x\rangle = -\mu\langle S_x\rangle, \tag{8.6}$$

where μ is the absorption coefficient, **S** is the Poynting vector $\mathbf{S} = (c/4\pi)\,\mathbf{E}\times\mathbf{H}$, and brackets denote time average. Since

$$\langle S_x\rangle = \frac{cE_{0y}H_{0z}}{4\pi}e^{-2\omega kx/c} \tag{8.7}$$

we have

$$\mu = \frac{2\omega k}{c} = \frac{4\pi\sigma}{nc}. \tag{8.8}$$

Although this derivation of Eq. (8.8) is the most direct, quantum-mechanical treatments of the absorption problem have frequently been set up in terms of the gradient of the energy density in the medium. It is therefore of interest to approach Eq. (8.8) in the following way. Let

$$\mu = \left(\frac{\text{Energy loss per unit volume per unit time}}{\text{Energy flux}}\right) = \frac{\left(\text{Power dissipation per unit volume}\right)}{\left(\text{Energy density}\right)\times\left(\text{Energy velocity}\right)}. \tag{8.9}$$

The total energy density in the electromagnetic field, including all energies stored in the medium, is [192]

$$\frac{\epsilon_1\langle \mathbf{E}^2\rangle}{4\pi} \tag{8.10}$$

where ϵ_1 must be regarded not as the dielectric constant but as a proportionality constant. It reduces to the ordinary dielectric constant only in regions of the spectrum where no appreciable absorption takes place. The power dissipated per unit volume is given by $\langle \mathbf{J}\cdot\mathbf{E}\rangle$ or $\sigma\langle \mathbf{E}^2\rangle$, and the velocity of energy propagation will be denoted by U_1. Then

$$\mu = \frac{\sigma\langle \mathbf{E}^2\rangle}{\frac{\epsilon_1}{4\pi}\langle \mathbf{E}^2\rangle U_1} = \frac{4\pi\sigma}{\epsilon_1 U_1}. \tag{8.11}$$

---

[192] L. Brillouin, *Congr. Intern. d'Élec. Paris* **2**, 739 (1932); "Wave Propagation and Group Velocity," Chapters 4 and 5. Academic Press, New York, 1959.

To compute ϵ_1 and U_1 separately, a detailed atomic picture of the medium is required. But it can be shown directly by flux continuity arguments[192] that even in regions of strong absorption, the product $\epsilon_1 U_1$ is always equal to nc, where n is the real part of the refractive index. Thus Eq. (8.11) is in fact equivalent to (8.8).

b. The Lorentz Model

In the Lorentz model, a dispersive and absorbing medium is characterized as a collection of harmonic oscillators of mass m, charge e, and natural (undamped) frequency ω_0, in the presence of a damping force. Each oscillator's equation of motion is

$$m(\ddot{\mathbf{u}} + \gamma \dot{\mathbf{u}} + \omega_0^2 \mathbf{u}) = \mathbf{F} = e\mathbf{E}, \tag{8.12}$$

where $\mathbf{u}$ measures its displacement from equilibrium and $\mathbf{F}$ is an applied force, generally an electrostatic force $e\mathbf{E}$. In the presence of an imposed electric field of the form (8.2a), a solution of Eq. (8.12) is

$$\mathbf{u} = \frac{e\mathbf{E}}{m}\left(\frac{1}{\omega_0^2 + i\omega\gamma - \omega^2}\right). \tag{8.13}$$

The contribution of one oscillator to the polarization density in the medium is taken to be $e\mathbf{u}$ (an appropriate distribution of positive charge is assumed present to maintain neutrality), and if the density of oscillators is n_0, we find that the complex dielectric displacement is

$$\mathbf{D} = \mathbf{E} + 4\pi \mathbf{P} = \mathbf{E} + 4\pi n_0 e\mathbf{u} = \left[1 + \frac{\omega_P^2}{\omega_0^2 + i\omega\gamma - \omega^2}\right]\mathbf{E}, \tag{8.14}$$

where $\omega_P^2 = 4\pi n_0 e^2/m$. The quantity in square brackets is just the complex dielectric constant N^2 introduced in Section 8a, from which we can obtain the ordinary dielectric constant and conductivity by using Eq. (8.4). Thus

$$\epsilon = \mathrm{Re}\, N^2 = 1 + \frac{\omega_P^2(\omega_0^2 - \omega^2)}{(\omega_0^2 - \omega^2)^2 + \omega^2\gamma^2} \tag{8.15}$$

and

$$\sigma = \frac{\omega_P^2}{4\pi}\frac{\omega^2\gamma}{(\omega_0^2 - \omega^2)^2 + \omega^2\gamma^2}, \tag{8.16}$$

8. CLASSICAL THEORY OF THE OPTICAL EFFECTS

which yield an absorption coefficient

$$\mu = \frac{4\pi\sigma}{nc} = \frac{\gamma}{n(\omega)\,c}\frac{\omega_p^2\omega^2}{(\omega_0^2 - \omega^2)^2 + \omega^2\gamma^2}. \tag{8.17}$$

Although it is possible to solve Eq. (8.5) to obtain an explicit functional form of $n(\omega)$, it is rather complicated and no explicit substitution will be made here. In spite of the fact that $n(\omega)$ is rapidly varying in the region of strong absorption, its variation is generally ignored and one speaks of "Lorentzian" forms of absorption curves. The second factor on the right in Eq. (8.17) is the Lorentzian shape. Actually, it is $n\mu$ and $n\omega k$ which have this shape.

Real solids have many fundamental excitations, and this fact is usually accounted for on the Lorentz model by the assumption that there exist many classical oscillators j at each lattice site, having different charges $ef_j^{1/2}$, natural frequencies ω_{0j}, and damping constants γ_j. Here f_j is the "strength" of the oscillator. Then

$$\epsilon = 1 + \sum_j \frac{f_j\omega_p^2(\omega_{0j}^2 - \omega^2)}{(\omega_{0j}^2 - \omega^2)^2 + \omega^2\gamma_j^2}, \tag{8.18}$$

with analogous expressions for σ and μ. This is not a completely satisfactory procedure for overlapping absorption bands, because the Lorentz model does not account properly for the effect of superposition of quantum-mechanical states, but it provides a good general picture of the optical properties.

When ω is not too far removed from a single isolated resonant frequency ω_{0k}, it is useful to set

$$\omega_{0k}^2 - \omega^2 = (\omega_{0k} + \omega)(\omega_{0k} - \omega) \approx 2\omega_{0k}(\omega_{0k} - \omega),$$

and thus make explicit the approximate symmetry of the Lorentzian factor:

$$\mu \text{ (near } \omega_{0k}) \approx \frac{\tfrac{1}{4}\gamma_k}{n(\omega)\,c}\frac{\omega_p^2}{(\omega_{0k} - \omega)^2 + \tfrac{1}{4}\gamma_k^2}. \tag{8.19}$$

c. "Retardation Effects"[193]

A closer inspection of certain results of the Lorentz model is useful

---

[193] This section closely parallels Sections 8 and 10 of M. Born and K. Huang, "Dynamical Theory of Crystal Lattices." Oxford Univ. Press, London and New York, 1954.

in connection with many recent papers on "retardation effects" on excitons. Let us ignore damping for the moment and apply the third Maxwell equation, (8.1c), to (8.14). We obtain

$$\left[1 + \frac{\omega_P^2}{\omega_0^2 - \omega^2}\right](\varkappa \cdot \mathbf{E}) = 0, \tag{8.20}$$

which may be satisfied in one of two ways: either the quantity in square brackets is zero, or $\varkappa \cdot \mathbf{E} = 0$. In the first case (Born and Huang's case A), it is readily shown that $\varkappa$, $\mathbf{E}$, and $\mathbf{u}$ (the oscillator displacement) must be parallel to one another, so that we are dealing with longitudinal waves. The longitudinal frequency ω_L is given by the value of ω which makes the square bracket in Eq. (8.20) vanish[194]:

$$\omega_L^2 = \omega_0^2 \left[1 + (\omega_P/\omega_0)^2\right]. \tag{8.21}$$

Case B, when $\varkappa \cdot \mathbf{E} = 0$, is the "transverse" case in which $\mathbf{E}$ and $\mathbf{u}$ are perpendicular to $\varkappa$. This is the case handled in Section 8b. Using Eq. (8.15) and recalling the definition of $N (\equiv c\kappa/\omega)$, we obtain, again in the case of no damping,

$$\frac{c^2\kappa^2}{\omega^2} = 1 + \frac{\omega_P^2}{\omega_0^2 - \omega^2}. \tag{8.22}$$

In contrast to the case of longitudinal solutions, the possible transverse frequencies $\omega(\kappa)$ are found to depend on κ, and in a complicated way. Their dependence is shown in Fig. 21.

The physical interpretation of the foregoing results may be taken over directly from the theory of lattice vibrations in ionic crystals. A large set of Lorentz oscillators of natural frequency ω_0 in a regular array will have two distinct kinds of propagating normal modes at long wavelengths.[195] One is longitudinal in the sense that vibration

---

[194] The square bracket is, of course, the dielectric function of the system and we are witnessing an example of the more general fact that longitudinal and transverse excitations of a system occur, respectively, at the zeroes and poles of its dielectric function; see J. Hubbard, *Proc. Phys. Soc. (London)* **A68**, 441 (1955).

[195] Although the present discussion assumes no coupling between the Lorentz oscillators themselves, no generality is lost because in the limit of very long wavelengths (very small κ) the coupling merely introduces an energy shift. A small effect, that of "spatial dispersion," will be discussed in Section 11. The normal modes of a lattice of Lorentz oscillators with dipolar coupling have been studied by Fano.[46]

takes place parallel to the direction of propagation, and the other is transverse. In the longitudinal case, each oscillator experiences an electric field (due to all the others) parallel to the direction of its motion. Its force constant is thereby effectively raised, and its fre-

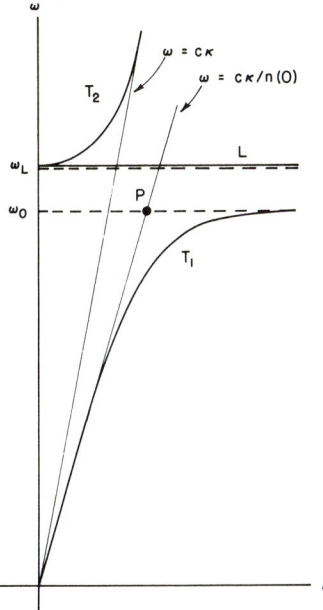

FIG. 21. "Vibrational frequencies" of the possible classical states of a system of Lorentz oscillators coupled to an electromagnetic field, as a function of wave vector. The dashed lines represent solutions in the absence of "retardation," i.e., in the absence of coupling between the field and the particles. The line of slope c is the pure photon dispersion curve outside the crystal while the line of slope $c/n(0)$ is the dispersion curve for a low frequency photon in the crystal. Others are discussed more fully in the text.

quency becomes ω_L ($> \omega_0$). In the transverse case, if no external electromagnetic field is present, there is no such electric field, and the natural frequency is undisturbed.[193] These two kinds of modes are pictured as horizontal lines at frequencies ω_L and ω_0 in Fig. 21. The line of slope c represents a photon in vacuum, and the range of wave vector κ in the figure is a very small fraction of the linear dimensions of a Brillouin zone of a crystal. Now it is clear that if the oscillators were vibrating in a transverse mode lying near the point P, they would like to radiate, and, conversely, the photons radiated would likely be absorbed. This fierce competition is indicative of the fact that *neither* photons *nor* transverse oscillator modes alone are good normal modes of the entire system (consisting of the crystal and radiation field) in this long wavelength region. The modes whose curves appear to cross at P actually repel each other and two mixed modes appear (T_1 and T_2). Their frequencies are given by the solutions of Eq. (8.22). At large values of κ, the resonance between photons and oscillators vanishes, and the pure photons and pure mechanical oscillations again become good normal modes of the system, as can be seen by the asymptotes of T_1 and T_2 toward the right.[196] At small κ, mixing is still strong. The lower branch (T_1) at $\kappa = 0$, for example, corresponds to a constant applied field; the Lorentz oscillators are polarized and thus "mix" with a photon of infinite wavelength. An interpretation of the absorption process along the following lines is thus suggested. When a steady flux of electromagnetic radiation is absorbed by a dielectric and equilibrium has been attained, a certain population of the true transverse normal modes must be maintained; these modes decay because their mechanical part is damped (coupled to an energy sink), and their population is constantly replenished at the expense of energy from the incident beam. While this interpretation is at variance with the popular notion of a photon being converted directly into a mechanical oscillation and then lost to an energy sink, it does not change the basic predictions of the Lorentz model. In fact, this interpretation has always been part of the Lorentz model but has been made more explicit in the analogous case of optical lattice vibrations.[50,197] The term "retardation effect" enters the theory in a somewhat indirect manner. If the Coulomb interaction is assumed to be instantaneous, which can be accomplished by setting $c = \infty$ (thus disregarding the time-dependent parts of Maxwell's equations), the possibility of radiation is ignored and the computation of the normal mode frequencies of the system of oscillators leads to the constant frequencies ω_0 and ω_L. A proper solution of the problem, the usual solution, *does* take the full Maxwell

---

[196] The asymptote of curve T_2 is shown with slope c in Figure 21. However, if there are other excitations (Lorentz oscillators) of the system at higher energies, they will reduce this slope to c/n', where n' is the index of refraction determined by these oscillators.

[197] K. B. Tolpygo, *Zh. Eksperim. i Teor. Fiz.* **20**, 497 (1950); K. Huang, *Proc. Roy. Soc.* **A208**, 352 (1951).

equations into account and the Coulomb interaction propagates with a finite velocity. Hence, sometimes there is said to be a "retardation effect."

d. Experimental Determination of the Optical Parameters

For convenience we record here some of the details of some common methods of direct measurement of the optical parameters. The absorption coefficient μ can be obtained directly from a measurement of an incident and transmitted intensity of a beam using a layer of thickness d (assumed homogeneous),

$$I_{\text{trans.}} = I_{\text{incident}}\, e^{-\mu d}. \tag{8.23}$$

A quantity proportional to μd is often called the "optical density" and is written

$$\text{O.D.} = \log_{10}(I_{\text{inc.}}/I_{\text{trans.}}) = \mu d/2.303. \tag{8.24}$$

Here I_{inc} must be the actual intensity just inside the first surface and I_{trans} that just inside the second surface. Thus, measurements of the O.D. must involve corrections for reflection, whenever the surfaces just mentioned are real ones.[198] Moreover, the presence of real surfaces makes it difficult to interpret optical phenomena as bulk processes when absorption coefficients are high. One technique which supposedly eliminates these difficulties is that of differential absorption[199]; the apparent O.D.'s of two films of different thicknesses (d_1, $d_2 > d_1$) are measured; the difference in O.D.'s ideally is independent of surface corrections and equal to the true O.D. of a film of thickness $d_2 - d_1$. "Ideal" conditions involve absence of interference effects, perfect equivalence of the two planes involved, etc.

A second means of obtaining optical data is from the reflectivity itself, which requires fewer corrections and which can be used in single crystals in regions where μd is to large to measure directly. The disadvantage of reflectivity measurements is the complicated dependence of the reflectivity on the complex index. In the case of a

---

[198] See, e.g., T. P. McLean, in "Progress in Semiconductors" (A. F. Gibson et al., eds.), Vol. 5, Sect. 3. Wiley, New York, 1961.

[199] See, e.g., J. E. Eby, K. J. Teegarden, and D. B. Dutton, Phys. Rev. 116, 1099 (1959).

plane wave at normal incidence, the ratio of the amplitudes of the reflected and incident waves is

$$r = \frac{n - ik - 1}{n - ik + 1} = \left[\frac{(n-1)^2 + k^2}{(n+1)^2 + k^2}\right] e^{i\theta(n,k)}. \tag{8.25}$$

The absolute value of r is measured directly, and the phase $\theta(n, k)$ is obtained from $|r|$ by use of a dispersion relation. Then n and k are deduced from $|r|$ and θ. Although $|r|$ is not known over the entire frequency range, reasonable assumptions may be made about its behavior at frequencies far removed from the one at which θ is being evaluated.[200]

9. Semiclassical Theory of Optical Absorption

By analogy with atomic processes, in which photons disappear and excitations of atoms are produced, the absorption of light by pure crystals in the visible and ultraviolet region is usually thought of in terms of the disappearance of a photon and the appearance of an exciton (or an electron-hole pair, which may be regarded as a continuum state of an exciton). But it has long been known that true absorption in a perfect crystal cannot take place by the creation of excitons alone. Were this process started, any population of excitons built up would soon decay back into photons of the same energy and momentum as those which were supposedly being absorbed. A sink for the energy stored in the exciton must be present, to compete with the radiative decay process, and in crystals containing relatively few localized defects this sink is probably provided in one way or another by lattice vibrations. Lattice vibrations are also responsible for line broadening, and the simultaneous interaction of excitons, phonons, and photons must be studied to understand absorption line shapes. In spite of this, the elementary transition

$$\text{(photon, no exciton)} \rightarrow \text{(exciton, no photon)} \tag{9.1}$$

in a system comprising the perfect crystal and a radiation field is of considerable importance, because its probability almost always

---

[200] For applications and discussions of this method see F. C. Jahoda, *Phys. Rev.* **107**, 1261 (1957) (BaO); H. R. Philipp and E. A. Taft, *Phys. Rev.* **113**, 1002 (1959) (Ge); M. P. Rimmer and D. L. Dexter, *J. Appl. Phys.* **31**, 775 (1960) (Ge).

9. SEMICLASSICAL THEORY OF OPTICAL ABSORPTION

governs the over-all strength of an exciton absorption line. In the semiclassical theory, the states of the photon-exciton-phonon system are tacitly assumed to be separable in the three field variables, and a matrix element of the photon-exciton interaction is seen to enter as a separate factor determining the transition probability for absorption from the ground state to any final state involving an exciton. Changes in the phonon population are in fact ignored, and no information on line broadening can be obtained from the theory. As we will see in Section 9c, the states indicated in Eq. (9.1) are not always true eigenstates of the whole system, just as in the case of Lorentz oscillators and photons. The photon-exciton coupling matrix element plays a central role in determining the true states when "retardation effects" are made explicit. The effects of phonons are studied in Section 10.

a. The Exciton-Photon Interaction

The traditional method of computing absorption coefficients of crystals involves the use of time-dependent perturbation theory. A radiation field with vector potential **A** incident on the crystal produces a perturbation on $\mathscr{H}_0$ (the Hamiltonian of Part II) such that the total Hamiltonian becomes[201]

$$\mathscr{H} = \mathscr{H}_0 - (e/mc) \sum_j \mathbf{A}(\mathbf{r}_j, t) \cdot \mathbf{p}_j + O(A^2) \qquad (9.2)$$

where $\mathbf{p}_j = -i\hbar \nabla_j$. The sum on j runs over the valence electrons referred to in the simplified model of Section 2, or, in a real crystal, over all electrons. Terms of order A^2 are neglected, as are interactions between the electromagnetic field and the cores; we will consider frequencies sufficiently high that phnon excitation by the photons is negligible. A vector potential which represents the electro-magnetic field introduced in Section 8 is

$$\mathbf{A}(\mathbf{r}, t) = \mathbf{A}_0 e^{i(\mathbf{\kappa} \cdot \mathbf{r} - \omega t)} + \text{c.c.} \qquad (9.3)$$

Although this potential should be damped, say by taking $\varkappa$ to be complex at the outset, its amplitude is generally assumed constant throughout "the crystal," which is interpreted to mean a region large

---

[201] L. I. Schiff, "Quantum Mechanics," 2nd ed., Chapter X. McGraw-Hill, New York, 1955.

enough to accomodate coherent excitons but small enough to cause negligible reduction of the amplitude of the field as absorption occurs. These two conditions are probably compatible only near weak absorption lines, but the results of this section are usually taken to be valid even in strongly absorbing regions of the spectrum.

The details of the computation of an absorption coefficient along the lines sketched above are well known for atoms[201] and have been written out frequently for excitons. We therefore give only the principal results. The probability per unit time that the crystal, in its ground state at $t = 0$, will have made a transition to a state $\Psi_{\nu\mathbf{K}}$ under the influence of the second term of Eq. (9.2) is proportional to $|H'_{\nu\mathbf{K},0}|^2$, where

$$H'_{\nu\mathbf{K},0} = \int \Psi^*_{\nu\mathbf{K}} \left[-(e/mc) \mathbf{A}_0 \cdot \sum_j e^{i\mathbf{\varkappa}\cdot\mathbf{r}_j}\mathbf{p}_j \right] \Psi_0 \, d\tau_1 \cdots d\tau_{2N}. \quad (9.4)$$

Because Ψ_0 is always assumed totally symmetric under the operations of the space group of the crystal, this matrix element will be zero unless the perturbation (in square brackets) and the state $\Psi_{\nu\mathbf{K}}$ have a common symmetry type. This means, first of all, that $\varkappa$ and $\mathbf{K}$ must be equal or must differ only by 2π times a reciprocal lattice vector; in practice, they are equal. Next, since generally some definite symmetry type has been settled on for the state $\Psi_{\nu\mathbf{K}}$, we look at the perturbation and ascertain its symmetry behavior by observing its transformation under the elements of the group of the $\mathbf{K}$ under consideration. Clearly the perturbation contains many different symmetry types, since we may write it

$$-(e/mc) \sum_j \mathbf{A}_0(\xi, \varkappa) \cdot \mathbf{p}_j [1 + i\varkappa \cdot \mathbf{r}_j + \cdots] \quad (9.5)$$

where now the precise nature of $\mathbf{A}_0$ is indicated by the associated wave vector $\varkappa$ and polarization index ($\xi = 1, 2$). The first term in Eq. (9.5) is responsible for dipole transitions, since it transforms like a vector in the direction of polarization, and the second is responsible for quadrupole transitions since it transforms like certain parts of the tensor $\Sigma_j \mathbf{p}_j \mathbf{r}_j$. These facts are well-known from atomic spectra, but here we must use care in their interpretation. If in a cubic crystal it has been determined that an exciton has Γ^-_{15} (p-like) symmetry at $\mathbf{K}=0$, we know from compatibility that in a 100 direction there will exist corresponding states of Δ_1 and Δ_5 symmetry. The polarization vector of light traveling in this direction transforms as Δ_5 and can only excite the Δ_5 ("transverse") states, as pointed out qualitatively in Section 3. At a general point in **K** space there is only one possible symmetry, and there is no strict polarization selection rule. Thus, if **K** lies *near* a symmetry line, one may be able to excite a state which is "mostly longitudinal," i.e., one which merges into a purely longitudinal state (such as Δ_1) as **K** moves to the symmetry line. These remarks apply only to transitions in which the dipole term of Eq. (9.5) makes the dominant contribution. Quadrupole[202] and magnetic dipole[203] transitions are just as well discussed in terms of the irreducible representations concerned.

We now look at Eq. (9.4) in more detail. Using Eq. (9.3) and the general exciton state (2.30), the matrix element (9.4) becomes

$$N^{-1/2} \sum_{\mathbf{R}} e^{-i\mathbf{K}\cdot\mathbf{R}} (-e/mc) \mathbf{A}_0(\xi, \varkappa)$$
$$\times \int \sum_{\beta} U(\beta)^* \Phi(\mathbf{R}, \mathbf{R}+\beta)^* \sum_j e^{i\varkappa\cdot\mathbf{r}_j} \mathbf{p}_j \Psi_0 \, d\tau_1 \cdots d\tau_{2N} . \quad (9.6)$$

Since only spin-independent, one-electron operators are involved, Eq. (9.6) vanishes unless the excited state contains some pure singlet component; we compute (9.6) for the case in which Φ and Ψ_0 are simple products of orthogonal functions. The integral can be rendered independent of **R** by the transformation of integration variables $\mathbf{r}_j \to \mathbf{R} + \mathbf{r}_j$, and we finally obtain [using Eq. (4.41)]

$$\left[N^{-1/2} \sum_{\mathbf{R}} e^{i(\mathbf{K}-\mathbf{K})\cdot\mathbf{R}} \right] (-e/mc) A_0(\xi, \varkappa) \, \mathbf{e}_{\xi\varkappa} \cdot (e^{i\varkappa\cdot\mathbf{r}} \mathbf{p})_{v0} \quad (9.7)$$

where, in the notation of Section 4,

$$(e^{i\varkappa\cdot\mathbf{r}} \mathbf{p})_{v0} \equiv \int \psi_{nv}^*(\mathbf{r}) \, e^{i\varkappa\cdot\mathbf{r}} \mathbf{p} a_{m0}(\mathbf{r}) \, d\tau \quad (9.8)$$

and we have introduced a unit polarization vector $\mathbf{e}_{\xi\varkappa}$ parallel to $\mathbf{A}_0(\xi, \varkappa)$. The factor in square brackets in Eq. (9.7) is just

---

[202] See, e.g., V. S. Galishev, V. I. Cherepanov, and R. V. Radchenko, *Fiz. Tverd. Tela* 3, 484 (1961); see *Soviet Phys.—Solid State (English Transl.)* 3, 355 (1961).

[203] A. G. Zhilich, V. I. Cherepanov, and Yu. A. Kargapolov, *Fiz. Tverd. Tela* 3, 1812 (1961); see *Soviet Phys.—Solid State (English Transl.)* 3, 1317 (1962).

$N\bar{\delta}(\varkappa - \mathbf{K})$,[204] and the remaining factors are precisely analogous to those which one obtains in the elementary theory of absorption of light by atoms. In the one-electron integral (9.8), a_{m0} represents the initial state of "an electron" which makes a transition and $\psi_{n\nu}$ represents its final state. The matrix element (9.8) is a useful special case of the slightly more general matrix element which appears as an integral in Eq. (9.6). For example, had we used the full singlet wave function for Φ instead of a simple product wave function, a factor $2^{1/2}$ would have entered into Eq. (9.8), and would later appear squared as the number of electrons available to make the particular one-electron transition under consideration. We are computing, therefore, "oscillator strengths per electron."

Finally, the square of Eq. (9.8) is

$$N \frac{e^2}{m^2 c^2} |\mathbf{A}_0|^2 \bar{\delta}(\varkappa - \mathbf{K}) |\mathbf{e}_{\xi\kappa} \cdot (e^{i\boldsymbol{\kappa}\cdot\mathbf{r}} \mathbf{p})_{\nu 0}|^2, \tag{9.9}$$

and the probability per unit time of the transition $\Phi_0 \to \Psi_{\nu\mathbf{K}}$ in the presence of radiation continuously distributed with a uniform spectral intensity $I(\omega)$ [in a range $d\omega$, $I(\omega)\, d\omega = (n\omega^2/2\pi c) |\mathbf{A}_0|^2$] is

$$dw = (4\pi^2 e^2/m^2 nc\hbar^2 \omega^2) I(\omega) N\bar{\delta}(\varkappa - \mathbf{K}) |\mathbf{e}_{\xi\kappa} \cdot (e^{i\boldsymbol{\kappa}\cdot\mathbf{r}} \mathbf{p})_{\nu 0}|^2 \delta(\omega - \omega_{\nu 0})\, d\omega, \tag{9.10}$$

where $\omega_{\nu 0} = (E_{\nu\mathbf{K}} - E_0)/\hbar$ and n is the index of refraction. The delta function in ω reminds us of the fact that this form of the theory is reliable only for the total strength of a given line. Only the *integrated* transition probability per unit time is independent of time. Now by Eq. (8.9), the absorption coefficient for radiation at frequency ω is

$$\mu = \frac{dw \cdot \hbar\omega \cdot V^{-1}}{I(\omega)\, d\omega} = \frac{4\pi e^2}{m^2 nc\hbar\omega} n_0 \bar{\delta}(\varkappa - \mathbf{K}) \delta(\omega - \omega_{\nu 0}) |\mathbf{e}_{\xi\kappa} \cdot (e^{i\boldsymbol{\kappa}\cdot\mathbf{r}} \mathbf{p})_{\nu 0}|^2, \tag{9.11}$$

where $n_0 \equiv N/V$ is the density of lattice points. The integrated absorption coefficient, with dimensions of energy $\times$ (length)$^{-1}$, is

$$\int \mu\, dE = \hbar \int \mu\, d\omega = \frac{4\pi^2 e^2}{m^2 nc\omega_{\nu 0}} n_0 |\mathbf{e}_{\xi\kappa} \cdot (e^{i\boldsymbol{\kappa}\cdot\mathbf{r}} \mathbf{p})_{\nu 0}|^2. \tag{9.12}$$

---

[204] The symbol $\bar{\delta}(\mathbf{k})$ will stand for a "lattice delta function,"

$$\bar{\delta}(\mathbf{k}) \equiv \Sigma_i\, \delta(\mathbf{k} - \mathbf{K}_i),$$

where the sum runs over all reciprocal lattice vectors.

9. SEMICLASSICAL THEORY OF OPTICAL ABSORPTION

The two delta functions in Eq. (9.11) represent conservation of momentum and energy, respectively, for the crystal-photon system. For photons of energy E (measured in ev), in a medium of index n, we have $\kappa \sim (5 \times 10^5)\, nE$ cm^{-1}, which is so near the origin of a typical BZ in a solid for visible and uv light that it has become traditional to speak of the creation of "$\mathbf{K} = 0$ excitions" by light. This is a valid approximation in many cases, but as we have seen in Section 6, important details may be overlooked if it is taken literally. The final factor in Eq. (9.11) governs the over-all transition probability once energy and momentum conservation are satisfied. When $\mathbf{p}$ itself has sizeable matrix elements between the hole and electron wave functions, we set $e^{i\kappa \cdot \mathbf{r}} \approx 1$ and call this the "dipole, allowed" case. Absorption can take place when $(\mathbf{p})_{\nu 0}$ has a projection on the polarization vector of the light.

In the preceding discussion, it has been assumed that the electric field at any point in the solid is equal to the externally applied (macroscopic) field, i.e., no "local field" effects have been included. This point is discussed below [Section 9b(1)].

b. Oscillator Strengths for Direct Transitions

The absorption coefficients computed above involve components of a dimensionless tensor

$$\mathbf{f}_{\nu 0} = \frac{2}{m(E_{\nu \mathbf{K}} - E_0)} (e^{i\kappa \cdot \mathbf{r}} \mathbf{p})_{0\nu} (e^{i\kappa \cdot \mathbf{r}} \mathbf{p})_{\nu 0}, \qquad (9.13)$$

the "oscillator strength" of the transition $\Phi_0 \to \Psi_{\nu \mathbf{K}}$. For light polarized parallel to the z axis, for example, Eq. (9.12) becomes

$$\int \mu dE = \frac{2\pi^2 e^2 \hbar}{mnc} n_0 f_{\nu 0}, \qquad (9.14)$$

where $f_{\nu 0}$ is the zz component of $f_{\nu 0}$. If the unit cell volume Ω_0 is measured in cubic angstroms,

$$\int \mu dE = 110 \times 10^6 \, (f_{\nu 0}/n\Omega_0) \text{ ev cm}^{-1}. \qquad (9.15)$$

In computations of oscillator strengths, it is usually very convenient to replace the matrix element of momentum between two states by a suitable multiple of the dipole matrix element between these two states.

Thus, we will often make the replacement

$$\mathbf{p}_{\nu 0} = (im/\hbar)(E_{\nu \mathbf{K}} - E_0)\,\mathbf{r}_{\nu 0}, \tag{9.16}$$

with the understanding that it is an approximation, because the states Φ_0 and $\Psi_{\nu \mathbf{K}}$ involved are not *exact* eigenstates of the system. Furthermore, to reduce the general theorem (which relates matrix elements of $\Sigma_i\,\mathbf{p}_i$ to those of $\Sigma_i\,\mathbf{r}_i$) to Eq. (9.16), which involves one-electron matrix elements but energies of the whole crystal, an additional set of simplifying assumptions is needed. The use of Eq. (9.16) does not appear to lead to serious errors in atomic problems where approximate states are used. In the event that $e^{i\mathbf{\kappa}\cdot\mathbf{r}}$ in Eq. (9.13) may be approximated by unity, we thus arrive at the following more familiar relation for the oscillator strength tensor for dipole-allowed transitions:

$$\mathbf{f}_{\nu 0} = \frac{2m}{\hbar^2}(E_{\nu \mathbf{K}} - E_0)\,\mathbf{r}_{0\nu}\,\mathbf{r}_{\nu 0}. \tag{9.17}$$

(1) *Frenkel's Case.* Here $\psi_{n\nu}^*$, the "envelope" of electron functions, consists merely of a single, excited electron state at the origin, and Eq. (9.8) reduces to the matrix element of $e^{i\mathbf{\kappa}\cdot\mathbf{r}}\mathbf{p}$ between two atomic functions on a single atom. (In making the following remarks we assume no overlapping between any atomic functions on different atoms.) Thus, except for the appearance of the exciton energy in the denominator instead of an atomic excitation energy, our general expression for the oscillator strength is merely an atomic oscillator strength. As far as order of magnitude is concerned, then, the Frenkel model predicts integrated absorption equivalent to a "gas" of density n_0 of free atoms or molecules with their normal oscillator strengths. An important question in principle, however, is whether to use the effective field or the macroscopic field in computing the absorption strength. Dexter[41] treated this problem on the extreme tight-binding model, pointing out that the wave functions $\Psi_{\nu \mathbf{K}}$ (which are of zeroth order with respect to the correlation energy of the crystal, as discussed is Section 2) are not good enough to compute oscillator strengths to an accuracy consistent with that of the energy, which is first-order. The use of a Lorentz effective field can be shown to account partially for the oscillator strength as computed with first-order wave functions, but such a procedure is reliable only when $(4\pi n_0/3)\,\alpha_a \ll 1$, where α_a is the atomic polarizability, and it does not account for any exchange and overlap effects. One effect

9. SEMICLASSICAL THEORY OF OPTICAL ABSORPTION

of the use of first-order wave functions is to reduce the zero-order oscillator strength by a factor $(1 - AE_{\nu K}^{-1})$, where A is of the order of magnitude of the exciton band width. Hoffman[205] has also derived this correction using a second-quantization formalism due to Agranovich[206]; the method involves a model Hamiltonian in which exchange and overlap effects are neglected, but it may be useful qualitatively if $AE_{\nu K}^{-1}$ happens to be large. Its value in solid argon is about 0.05, and it will undoubtedly be small in most molecular crystals. Probably the most interesting aspect of the subject of oscillator strengths in molecular crystals is their polarization dependence in the presence of Davydov splitting. As remarked in Section 3, the true states $\Psi_{\nu K}$ for certain molecular crystals may be intricate mixtures of excitons constructed from similar but differently-oriented localized excited states. In this case, $|\mathbf{e}_{\xi K} \cdot \mathbf{p}_{\nu 0}|^2$ will be a function of the mixing coefficients and the relative angles between the polarization vector, the crystal axes, and the orientation vectors of the individual molecules in a unit cell. A general form of the theory is due to Winston,[58] and comparison with experimental data is reviewed by McClure.[14]

(2) *Wannier's Case.*[21,207] Elliott's extensive calculations for Wannier excitons are based mainly on the Bloch representation rather than the localized representation which we have used [e.g., in Eq. (9.8)]. For the sake of comparison with Frenkel and intermediate-case excitons, we retain the localized representation. Consider first the case of dipole-allowed transitions. The z component of $\mathbf{r}_{\nu 0}$ is

$$z_{\nu 0} = \int \psi_{n\nu}(\mathbf{r})^* z a_{m0}(\mathbf{r}) \, d\mathbf{r}$$

$$= \sum_{\beta} U_{\nu K}(\beta)^* \int a_{n\beta}(\mathbf{r})^* z a_{m0}(\mathbf{r}) \, d\mathbf{r}. \quad (9.18)$$

The sum includes one term in which the electron and hole are at the same lattice site ($\beta = 0$), and if the corresponding integral does not vanish, this term dominates the sum and all other terms may be

---

[205] R. Hoffman, *Vestn. Mosk. Univ., Ser. III: Fiz. Astron.*, p. 69 (1962).
[206] V. M. Agranovich, *Zh. Eksperim. i Teor. Fiz.* **37**, 430 (1959), see *Soviet Phys. J.E.T.P. (English Transl.)* **10**, 307 (1960).
[207] G. Dresselhaus, *Phys. Rev.* **106**, 76 (1957).

considered "two-center corrections." In that case, (9.18) becomes $U_{\nu K}(0)^* z_{nm}$, where z_{nm} is the integral

$$\int a_{n0}^*(\mathbf{r}) \, z a_{m0}(\mathbf{r}) \, d\mathbf{r}, \tag{9.19}$$

a quantity which is essentially an atomic dipole matrix element and which has that order of magnitude. The oscillator strength per electron is then

$$f_{\nu 0} = \frac{2m}{\hbar^2} (E_{\nu K} - E_0) \, | \, U_{\nu K}(0) \, |^2 \, | \, z_{nm} \, |^2. \tag{9.20}$$

The factor $| \, U_{\nu K}(0) \, |^2$ is the fraction of time which the electron spends at the lattice site of the hole, and is equal to unity on the Frenkel model. In the Wannier continuum approximation it is proportional to $| \, F_\nu(0) \, |^2$ and, since F_ν is a hydrogenic function, it is nonvanishing only for s-like envelope functions.

The vanishing of Eq. (9.19) because a_n and a_m are of the same parity, or are of the wrong symmetry otherwise, is not sufficient to preclude dipole-allowed transitions. If this one-center term vanishes, we may still look at the two-center terms in Eq. (9.18). For simplicity let a_m and a_n both be totally symmetric [in which case Eq. (9.19) certainly will have vanished] and evaluate (9.22). If $\mathbf{\beta}_0$ and $-\mathbf{\beta}_0$ are the positions, along the z axis, of the 2 neighbors nearest the origin,[208] then

$$z_{\nu 0} \approx [U_{\nu K}(\mathbf{\beta}_0) - U_{\nu K}(-\mathbf{\beta}_0)] \int a_{n\beta_0}(\mathbf{r})^* z a_{m0}(\mathbf{r}) \, d\mathbf{r}. \tag{9.21}$$

In obtaining Eq. (9.21) use has been made of the fact that the two-center matrix element of z differs only in sign for the two terms included in the summation. This result makes it clear that we may still obtain an allowed dipole transition if the *envelope* function is of odd parity. That is, Eq. (9.21) does not necessarily vanish if

$$U_{\nu K}(-\mathbf{\beta}_0) = -U_{\nu K}(\mathbf{\beta}_0),$$

which will be the case for a p-like envelope.

In the continuum approximation, the distance β_0 is considered small enough to be a differential, so we may make the replacement

$$\frac{U_{\nu K}(\mathbf{\beta}_0) - U_{\nu K}(-\mathbf{\beta}_0)}{2\beta_0} \to \frac{\partial U_{\nu K}(0)}{\partial \beta_z}. \tag{9.22}$$

---

[208] This discussion is based on a simple cubic crystal, but can be readily generalized.

9. SEMICLASSICAL THEORY OF OPTICAL ABSORPTION

The matrix element (9.21) is thus given by

$$z_{v0} \to 2\beta_0 z_{nm}(\beta_0) \frac{\partial U_{vK}(0)}{\partial \beta_z} \qquad (9.23)$$

in which $z_{nm}(\beta_0)$ is the two-center integral appearing in Eq. (9.21). Combining Eqs. (9.23) and (9.17), we obtain

$$f_{v0} = \frac{2m}{\hbar^2}(E_{vK} - E_0) \mid 2\beta_0 z_{nm}(\beta_0) \mid^2 \left| \frac{\partial U_{vK}(0)}{\partial \beta} \right|^2. \qquad (9.24)$$

Using the known envelope functions for hydrogenic excitons, we see that Eq. (9.20) is nonvanishing only for s-like envelopes, and (9.24) is nonvanishing only for p-like envelopes. Their actual values are

$$f_{v0}(\text{"allowed"}) = \frac{2m(E_{vK} - E_0)}{\hbar^2} \mid z_{nm} \mid^2 \frac{\Omega_0}{\pi a^3} \frac{1}{n^3} \qquad (9.25)$$

$$f_{v0}(\text{"forbidden"}) = \frac{2m(E_{vK} - E_0)}{\hbar^2} \mid 2\beta_0 z_{nm}(\beta_0) \mid^2 \frac{\Omega_0}{3\pi a^5} \frac{n^2 - 1}{n^5} \qquad (9.26)$$

where n is the principal hydrogenic quantum number, $a = \epsilon \hbar^2 / \mu e^2$ is the exciton Bohr radius, and Ω_0 is the unit cell volume (see the discussion of normalization of exciton wave functions in Section 4a). It is seen that "forbidden" transitions are weaker than "allowed" transitions by at least a factor $(\beta_0/a)^2$. Clearly they vanish entirely for $n = 1$, where there exists no p-like envelope.[209]

The strong n dependence of the oscillator strengths given by Eqs. (9.25) and (9.26) can be easily subjected to direct tests. Thus for an "allowed" series, the intensities should be in the approximate ratios $1 : \frac{1}{8} : \frac{1}{27}$ for $n = 1, 2, 3$, respectively; a ratio of $1/8.1$ between

---

[209] Since the "allowed" and "forbidden" cases are, as emphasized above, merely two different examples of dipole-allowed transitions, some authors[53] call them both *allowed*. We agree with these authors but prefer to compromise by using quotation marks around the names. "Forbidden" transitions are forbidden only in the sense that a one-center dipole matrix element vanishes (or, equivalently, transitions at precisely $\mathbf{k} = 0$ in the Bloch scheme are forbidden). They are allowed in all the usual senses: no phonons are required to make them go, and they represent single-photon absorption with photon-exciton momentum conservation. Nikitine [see, e.g., S. Nikitine, R. Reiss, and M. Sieskind, *Compt. Rend.* **246**, 3439 (1958)] calls absorption spectra unambiguously associated with (9.25) and (9.26) *first* and *second* class spectra, respectively; all other spectra are *third* class.

the $n = 1$ and 2 peaks in CuI has been found, [210] and a line 1/9 the strength of, and lying just above, the lowest absorption line has been found in an unlikely place, solid xenon.[64] For the "forbidden" series, the ratios are $0 : \frac{3}{32} : \frac{8}{243}$. The $n = 2, 3$ lines of the Cu_2O yellow[80] and green[211] series have strengths obeying this law. Grun and co-workers[211] have also compared the strengths of the $n = 2$ lines *between* the two series. The ratio of strengths should be, according to (9.26),

$$\frac{f_{2y}}{f_{2g}} = \frac{E_{2y}}{E_{2g}} \times \frac{a_g^5}{a_y^5} \times \frac{|z_y(\beta_0)|^2}{|z_g(\beta_0)|^2}. \qquad (9.27)$$

Here subscripts g and y are used to denote quantities associated with the green and yellow series, respectively. The value of the first two factors in Eq. (9.27), as deduced from the transition energies and binding energies of the two excitons,[212] assuming the same dielectric constant for both, is (cf. Section 4)

$$\frac{2.149}{2.267} \times \frac{0.0969}{0.1540} = \frac{1}{10.8}.$$

The experimental ratio is 1/11,[211] so the two-center matrix elements in Eq. (9.27) must be practically identical for both series. There seems to be no particular reason for this, since the matrix elements depend on the details of the Wannier functions of the particular valence bands involved.

The absolute values of oscillator strengths have been measured in many crystals by French workers, and again we select Cu_2O for a brief discussion. The $n = 2$ yellow line has a strength of $f_{2y} = 2.8 \times 10^{-6}$ at 4.2° K. Using Eq. (9.26) along with parameters derived from the spectrum, one may deduce a value of $|z_y(\beta_0)|$. We take β_0 as the Cu-Cu distance, $7 a_0$; $\Omega_0 = 26.2 a_0^3$, $E_{2y} - E_0 = 2.149$ ev, $a = \epsilon\mu^{-1} a_0 = 16.3 a_0$, and $n = 2$. The result is $|z_y(\beta_0)| = 0.55 a_0$, which is quite a reasonable value for a two-center dipole matrix element. Thus the theory is in fairly good quantitative agreement with the measured oscillator strenghts.

---

[210] See the discussion by S. Nikitine, *in* "Progress in Semiconductors" (A. F. Gibson *et al.*, eds.), Vol. 6, pp. 291-293. Wiley, New York, 1962.

[211] J. B. Grun, M. Sieskind, and S. Nikitine, *J. Phys. Radium* **22**, 176 (1961).

[212] For a fixed dielectric constant, the exciton Bohr radius, a_i, for a given series is inversely proportional to the exciton rydberg, G_i, for that series.

The oscillator strengths above refer to transitions from the crystal ground state to discrete levels. When the hydrogenic levels become so densely spaced that they resemble a continuum, a constant absorption coefficient results in the "allowed" case, because while the intensity per line is dropping off as n^{-3}, the density of lines is going up as n^3. In the true continuum, when the electron and hole are being sent into what are considered to be free states on the one-electron (band) model, but which are to be regarded as hydrogenic continuum states from the point of view of exciton theory, the computed absorption coefficient for photons of energy $E_{\nu K} - E_0 = \hbar\omega$ has the form[21]

$$\mu = \frac{4\pi\omega \, |z_{nm}|^2 \, \epsilon}{a^2 cn(\omega)} \frac{1}{1 - e^{-2\pi\alpha}} \qquad (9.28)$$

where $\alpha = (G/\Delta E)^{1/2}$ and $\Delta E = E_{\nu K} - E_0 - E_G$ (the energy of relative motion of the hole and electron, i.e., the energy in excess of the band gap energy E_G). Also, a is the exciton Bohr radius; ϵ is the dielectric constant used in determining the exciton energy; $G = \mu e^4/2\hbar^2\epsilon^2$ is the exciton rydberg; and $n(\omega)$ is the real part of the complex index. Equation (9.28) reduces to the constant deduced from the quasi-continuum just below the band edge, when $\Delta E \to 0$, and to the Bardeen-Blatt-Hall[213] shape $\mu \sim (\Delta E)^{1/2}$ when $\Delta E \gg G$. In the latter case, one would expect the results to approach those of simple band theory, because the kinetic energy of the relative motion of the electron and hole dominates the Coulomb energy. Analogous results are given by Elliott for the "forbidden" case; there, one obtains

$$\mu = \frac{4\pi\omega \, |2\beta_0 z_{nm}(\beta_0)|^2 \, \epsilon}{3a^4 cn(\omega)} \frac{1 + \alpha^{-2}}{1 - e^{-2\pi\alpha}}. \qquad (9.29)$$

This also reduces to the free-particle shape, $\mu \sim (\Delta E)^{3/2}$, when $\Delta E \gg G$, and to a value consistent with the quasi-continuum just below the band gap.

In the event that a dipole transition is truly forbidden by symmetry, which happens to be the case in the $1s$ line of the yellow series of Cu_2O, quadrupole absorption may be possible. In this case, the matrix element $(e^{i\kappa \cdot r} \mathbf{p})_{0\nu}$ is approximately equal to $(i\mathbf{\kappa} \cdot \mathbf{r} \, \mathbf{p})_{0\nu}$, and the transition probability contains the square of

$$\mathbf{\kappa} \cdot (\mathbf{rp})_{0\nu} \cdot \mathbf{e}_{\xi\kappa}. \qquad (9.30)$$

[213] J. Bardeen, F. Blatt, and L. H. Hall, in "Photoconductivity Conference," p. 146. (R. G. Breckenridge et al., eds.). Wiley, New York, 1956.

Although a second-rank Cartesian tensor like **rp** contains three irreducible parts, only the D_2 (d-like) part contributes to our result because $\mathbf{e}_{\xi\kappa}$ is perpendicular to $\varkappa$. Transitions may therefore take place to any exciton states whose over-all symmetries are compatible at Γ with those found in the list of representations subduced from D_2 in the given crystal. In the case of a cubic crystal, these symmetries are Γ_{12} and $\Gamma_{25}{'}$. Whether a low-lying exciton state of one of these types exists depends entirely on the crystal involved, and, in the case of a description on the Wannier model, on the conduction and valence band symmetries at their extrema. The two types (Γ_{12} and $\Gamma_{25}{'}$) may be distinguished from one another experimentally, however, by the details of the anisotropy introduced into the absorption spectrum. Using this fact, Elliott[70] has deduced that the Cu_2O $1s$ exciton has $\Gamma_{25}{'}$ symmetry and is probably built from a Γ_6 conduction band and a Γ_7 valence band. The $1s$ line is known to be a very weak line whose oscillator strength[214] is about 10^{-9}. Despite this, its behavior under strain[157] and magnetic fields[215] has been studied extensively. For a fuller discussion of the details of quadrupole transitions in strained and unstrained crystals, the reader is referred to the articles of Elliott and of Cherepanov and colleagues.[202,216]

Intensities of exciton absorption can be influenced strongly by magnetic fields under certain circumstances. We discuss here an effect apparently discovered independently by Gross's group[217] and by Hopfield and Thomas.[53] Using a field of 31 koe, the latter found pairs of lines in CdS whose intensities were interchanged on reversal of the magnetic field when the field, the direction of photon propagation, and the c axis were mutually perpendicular (see Fig. 22). With a fixed field the reversal did not occur when the crystal was rotated 180° about the c axis, but did occur when it was rotated by 180°

---

[214] E. F. Gross and A. A. Kaplianskii, *Dokl. Akad. Nauk S.S.S.R.* **139**, 75 (1961); see *Soviet Phys. "Doklady" (English Transl.)* **6**, 592 (1962).

[215] E. F. Gross, A. G. Zhilich, B. P. Zakharchenya, and A. V. Varfalomeev, *Fiz. Tverd. Tela* **3**, 1445 (1961); see *Soviet Phys.—Solid State (English Transl.)* **3**, 1048 (1961).

[216] V. I. Cherepanov and V. S. Galishev, *Fiz. Tverd. Tela* **3**, 1085 (1961); see *Soviet Phys.—Solid State (English Transl.)* **3**, 790 (1961); V. I. Cherepanov, V. V. Druzhinin, Yu. A. Kargapolov, and A. E. Nikiforov, *Fiz. Tverd. Tela* **3**, 2987 (1961); see *Soviet Phys.—Solid State (English Transl.)* **3**, 2179 (1962); V. I. Cherepanov, *ibid.* **3**, 1082, 1583, 1894, 2179 (1961, 1962).

[217] E. F. Gross, B. P. Zakharchenya, and O. V. Konstantinov, *Fiz. Tverd. Tela* **3**, 305 (1961); see *Soviet Phys.—Solid State (English Transl.)* **3**, 221 (1961).

about the y axis (direction of propagation). The latter is seen to be equivalent to reversing the direction of $\varkappa$, keeping **H** and the c axis fixed. These effects can be traced to the lack of inversion symmetry in CdS; for in such a case it is possible for the dipole and quadrupole

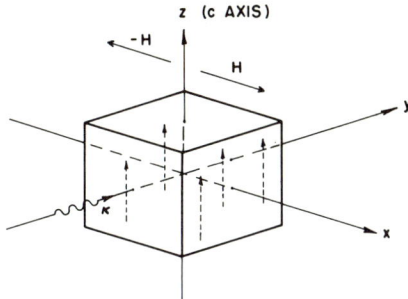

FIG. 22. Geometry of the CdS intensity-reversal experiment (see text). The dashed arrows represent the directionality of the c axis.

terms in Eq. (9.5) to contribute simultaneously to the transition matrix element. Thus the transition probability to a state will contain a factor

$$\left| \mathbf{e}_{\xi\kappa} \cdot \int \psi_{nv}^* \, \mathbf{p}(1 + i\varkappa \cdot \mathbf{r}) \, a_{m\mathbf{0}} \, d\tau \right|^2 \equiv |D_\nu + \kappa_y Q_\nu|^2 \quad (9.31)$$

for photons incident in the y direction. D_ν and Q_ν need not have a fixed relation to one another for all exciton states and indeed, in CdS, there are certain pairs of $2P_{xy}$ degenerate states to which relative transition probabilities are $|D + \kappa_y Q|^2$ and $|D - \kappa_y Q|^2$. On the application of the magnetic field the degeneracy of these states is raised, as sketched in Fig. 23, and the difference in transition prob-

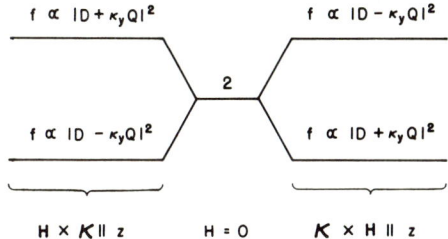

FIG. 23. Typical magnetic-field behavior of positions and strengths of absorption lines in the intensity-reversal experiment. Energy is plotted vertically.

abilities is exposed. Reversing the magnetic field maintains the splitting but interchanges the *states* involved, accounting for the intensity interchange mentioned first above. It is clear, too, from the form of the oscillator strengths, that the rotation which effectively reverses κ_y will also interchange the intensities. This discussion of the mixed dipole-quadrupole transitions has necessarily neglected many fine points, and the reader is enthusiastically referred to the complete analysis given by Hopfield and Thomas.[53] Degeneracies are generally higher than those indicated here; the coefficients D and Q are geometry-dependent and in many cases magnetic-field dependent, according to the extent to which the field affects the form of the eigenstates; some CdS absorption lines, because of degeneracies of states of different symmetries, are superpositions of pure dipole, mixed dipole-quadrupole, and pure quadrupole transitions. Careful analysis of the transition probabilities in various geometries has contributed significantly to the wealth of information now available on the band and exciton structure of CdS.

No matter how intriguing CdS may be in these respects, Thomas and Hopfield conclude reluctantly that the anomalous effects are accidental. The principal "accident" is that the quadrupole parts of the matrix elements are roughly of the same order of magnitude as the dipole parts. They discuss the conditions necessary for the effects to be observed in other crystals lacking inversion symmetry.

At very high magnetic fields special techniques are necessary. Elliott and Loudon[138,139] compute eigenvalues and eigenfunctions by means of the artificial separation of coordinates introduced for hydrogen by Schiff and Snyder.[218] The results are valid for fields sufficiently large that

$$\beta \equiv \epsilon^2 \hbar^3 H_z / \mu^2 e^3 c$$
$$= (e H_z \hbar / \mu c)/2G \gg 1 \qquad (9.32)$$

and consist of states described by a wave function in the relative coordinate which (in cylindrical coordinates) is a product of the form

$$F_\nu(\mathbf{r}) = \psi(\rho, \phi) f(z). \qquad (9.33)$$

Here $\psi(\rho, \phi)$ is the same function of ρ, ϕ as in the absence of the Coulomb interaction, and $f(z)$ is the solution of an effective Schrödinger equation in which a potential $V(z)$ appears as the average of $-e^2/\epsilon r$ over the ρ, ϕ coordinates. The magnetic field is in the z

---

[218] L. I. Schiff and H. Snyder, *Phys. Rev.* **55**, 59 (1939).

direction. In the absence of the Coulomb interaction, $f(z)$ is of course just a plane wave solution Ce^{ikz}. In the presence of the Coulomb interaction, $f(z)$ is the solution of a complicated differential equation which can be converted by reasonable approximations into Whittaker's equation. This done, there are seen to exist (a) bound solutions, and (b) continuum solutions which bear the same relationship to Landau states as do Coulomb continuum states to plane waves. The bound solutions are the high-field counterparts of exciton states, and transitions to the lowest one are relatively quite strong, as shown in Fig. 24. The curve marked B is the predicted absorption coefficient for an "allowed" hydrogenic series in which the lines have an assumed width of one exciton rydberg; thus only one line is discernible in the series. The curve marked D is the absorption curve for the same case, but with a magnetic field such that the parameter β [Eq. (9.32)] is equal to 2. There is seen to be a shift in the main peak and considerably

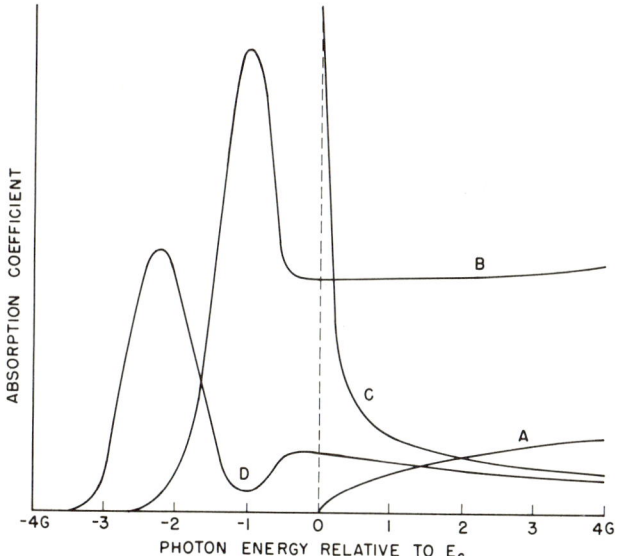

FIG. 24. Comparison of predictions of absorption edge shapes for the following modifications of simple band-to-band transitions. (A) No modification: Coulomb interaction and magnetic field absent. (B) Coulomb interaction only (normal exciton spectrum). (C) Magnetic field only. (D) Simultaneous effect of large magnetic field and Coulomb interaction. A width of one exciton rydberg G is assumed for the exciton lines; abscissa is in units of G. [After R. J. Elliott and R. Loudon, *Phys. Chem. Solids* 15, 196 (1960).]

less absorption in the continuum, but the more striking contrast is between these two cases and the corresponding results for simple valence and conduction bands with the Coulomb interaction neglected. Curve A is the absorption expected from the creation of free hole-electron pairs, and C shows the effect of the magnetic field on this curve.[219] For weakly bound excitons of moderate broadening, it is clear from curve B that the Coulomb interaction prevents any simple positioning of the band gap energy in an absorption spectrum, even in the absence of a magnetic field; in the presence of a strong field, the Coulomb interaction moves the free-pair peak (curve C) downward in energy by an amount of the order of a few exciton rydbergs in the case considered here.

Elliott and Loudon's calculations do not apply to any real crystal at present, since the only case where β has been made sufficiently large is in Ge, where the band structure is not simple. However, their results appear to be qualitatively correct,[149] and must be taken into consideration whenever an interpretation of magnetoabsorption data is attempted on the basis of free-particle Landau levels.

(3) *Intermediate Cases.* In the Frenkel case, both the "envelope" function and the relevant localized wave functions of the exciton are assumed known at the outset. In the Wannier case, the envelope is assumed to be given fairly accurately by a hydrogenic function, and most theoretical attention has been focused on general characteristics of the hydrogenic series (Elliott) and relative transition probabilities among states demonstrably derivable from the same conduction and valence bands (Thomas-Hopfield). Little attention has been paid to absolute calculations, i.e., calculations of z_{nm} and $z_{nm}(\beta_0)$ in Eqs. (9.25) and (9.26), because explicit one-electron functions are not even available in Bloch form for the crystals of major interest. In intermediate cases, *neither* the envelopes nor the localized functions are known with great accuracy, and the comparison of oscillator strengths with predictions of various models can provide a means of assessing the model wave functions.

The only work along these lines, to the author's knowledge, concerns NaCl.[10,11,122,123] As pointed out in Section 5, the orbital part of the hole wave function a_{m0} is taken as that of a Cl⁻ $3p$ electron, and the orbital part of the electron function ψ_{nv} is controversial, but

[219] B. Lax, L. M. Roth, and S. Zwerdling, *Phys. Chem. Solids* **8**, 311 (1959).

it is quite likely to be totally symmetric in the two lowest exciton states of interest. In the excitation model it is taken to be

$$\psi_E = Ce^{-r/\alpha} g(\mathbf{r}), \tag{9.34}$$

where $g(\mathbf{r})$ is the conduction band wave function at $\mathbf{k} = 0$ and α is a damping length of order 2.5 to 3 Bohr radii. In the transfer model it is essentially

$$\psi_T = (6O)^{-1/2} \sum_j \phi_{Na3s}(\mathbf{r} - \mathbf{R}_j) \tag{9.35}$$

where O is a normalization factor and Σ_j runs over the 6 Na$^+$ sites nearest neighboring the Cl$^-$ site at which the hole is located. The experimental oscillator strength per electron for transitions into the lowest doublet peak in NaCl is estimated[10,220] as 0.05. Using the electron wave functions (9.34) and (9.35) one obtains 0.06 and 0.22, respectively.[93] Thus the excitation model is the better of the two in this application. A more elaborate electron-transfer treatment[122] using linear combinations of determinantal wave functions yields 0.9. Although the transfer calculation using Eq. (9.35) indicates better agreement with experiment, it is less accurate than the calculations with determinantal wave functions because no simple overlap between ϕ_{Na3s} and the Cl$^-$ neighbors is admitted (O corrects only for $3s$-$3s$ overlap). Thus Eq. (9.35) omits modulation of the $3s$ function near the Cl nuclei. The determinantal functions, on the other hand, are known to be unreliable because they do not predict the transition energy at all well. Calculations of the oscillator strengths with the Dykman-Tsertsvadze wave functions[119] are not available.

An interesting feature of the oscillator strengths associated with exciton absorption in the alkali halides and rare gases is the occasional inversion of the intensity ratio expected in the lowest doublet. In analogous atomic absorption spectra, the corresponding doublet derives, at the jj coupling limit, from transitions to excited states labeled $(\tfrac{3}{2} \tfrac{1}{2})$ and $(\tfrac{1}{2} \tfrac{1}{2})$. When Hund's rule of singlet-triplet level ordering is valid, the transition to $(\tfrac{3}{2} \tfrac{1}{2})$ is of lower energy and is the stronger one.[221] Although Hund's rule does not necessarily hold for

---

[220] S. G. El Komoss, *Phys. Chem. Solids* **23**, 55 (1962).

[221] In the pure atomic jj limit, the $(\tfrac{3}{2} \tfrac{1}{2})$ line is twice as strong as $(\tfrac{1}{2} \tfrac{1}{2})$, the higher-energy component. There is a crossover of strengths as the LS limit is approached, roughly when spin-orbit and exchange energies are equal.

excitons,[128,222] it seems that the *jj* limit is being very closely approached in AgCl and AgBr,[223] RbBr and KBr,[109] and Kr.[64] Alternatively, these data may be evidence that the lowest direct exciton is not associated with electronic states at the center of the zone.[224]

It has been suggested[225] that the alkali halide doublet strength ratio should be the same as that in the related transitions in the rare gas atom analogous to the halide. This model is of quite restricted usefulness, since it assumes precisely the same degree of intermediate coupling in the two systems.

(4) *Miscellaneous Work*. Because space is limited, we cannot go into a detailed analysis of each of the absorption spectra measured in recent years. Most interpretations of these spectra, except in the case of the alkali halides, have tended to follow the Wannier model, because of its reasonably simple predictions. Good review articles are available,[16–18] and we list here several other solids in which both measurements and serious interpretation have been undertaken: HgI_2,[226] Cu_2O (higher conduction bands),[227] copper halides,[228] several IIA—IVB compounds,[229] LiF,[110] thallium halides,[230,231] and the rare gases.[64,67]

c. Limitations of the Method; "Retardation Effects"

As we have already indicated, this elementary theory of line

---

[222] R. S. Knox, *Progr. Theoret. Phys. (Kyoto)* **24**, 909 (1960).

[223] Y. Okamoto, *Nachr. Akad. Wiss. Goettingen, II. Math.-Physik. Kl.* No. 14, 275 (1959).

[224] This conjecture is prompted by conversations with Dr. H. Ehrenreich, who has considered it as a possible explanation of certain reflectivity data on certain alkali halides.

[225] R. Petersen, *J. Phys. Chem. Solids* **1**, 284 (1957).

[226] M. Sieskind, J. B. Grun, and S. Nikitine, *J. Phys. Radium* **22**, 51 (1961).

[227] J. B. Grun, M. Sieskind, and S. Nikitine, *Phys. Chem. Solids* **21**, 119 (1961).

[228] S. Nikitine, S. G. El Komoss, R. Reiss, and J. Ringeisen, *J. Chim. Phys.* **55**, 665 (1958); M. S. Brodin and A. S. Krochuk, *Ukr. Fiz. Zh.* **6**, 706 (1961); R. Coelho, *Proc. Intern. Conf. Semicond. (Czech. Acad. Sci., Prague, 1961)*, p. 686; S. Nikitine and R. Reiss, *J. Phys. Chem. Solids* **16**, 237 (1960); R. Reiss and S. Nikitine, *Compt. rend. acad. Sci.* **250**, 2862 (1960).

[229] R. J. Zollweg, *Phys. Rev.* **111**, 113 (1958); G. H. Reiling and E. B. Hensley, *Phys. Rev.* **112**, 1106 (1958); G. A. Saum and E. B. Hensley, *ibid.* **113**, 1019 (1959).

[230] H. Zinngrebe, *Z. Physik* **154**, 495 (1959); S. Tutihasi, *Phys. Chem. Solids* **12**, 344 (1960).

[231] S. Nikitine, J. Biellmann, and M. Sieskind, *J. Chim. Phys.* **55**, 664 (1958).

9. SEMICLASSICAL THEORY OF OPTICAL ABSORPTION

strengths does not provide a prediction of the shape of an absorption or dispersion curve, and it explains the absorption process itself only insofar as the conversion of a photon to an exciton is assumed to be followed by decay to some as yet unspecified nonradiative state of reasonably long lifetime. It is worth noting[232] that the transition probability (9.10) for absorption contains a factor N, which, although reasonable in terms of what we expect in absorption, also would appear in the spontaneous transition probability for emission. It arises from the assumption of complete coherence of the photon over the whole crystal, and implies an unphysically short lifetime ($\sim 10^{-30}$ sec) for the exciton state $\Psi_{\nu\mathbf{K}}$ coupled to the electromagnetic field. This paradox is probably best understood in terms of "retardation," the quantum analog of the Lorentz theory, on the basis of work by Fano[46] and Hopfield,[233] who point out that neither a single photon nor a single exciton is a good eigenstate of the Hamiltonian of the *whole* system consisting of the electromagnetic field and the crystal.

To properly set the present problem and for future reference, let us translate Section 9a into the language of second quantization. The crystal and photon field is described in zeroth order by functionals $|\cdots N_{\nu\mathbf{K}}\cdots, \cdots n_{\xi\mathbf{\kappa}}\cdots\rangle$, in which $\{N_{\nu\mathbf{K}}\}$ and $\{n_{\xi\mathbf{\kappa}}\}$ are sets of exciton and photon occupation numbers, respectively; $n_{\xi\mathbf{\kappa}}$ is the number of photons present with wave vector $\mathbf{\kappa}$ and polarization ξ. The Hamiltonian operator for the system (crystal + radiation) with a linear interaction is taken to be

$$H = \sum_{\nu\mathbf{K}} E_{\nu\mathbf{K}} b^\dagger_{\nu\mathbf{K}} b_{\nu\mathbf{K}} + \sum_{\xi\mathbf{\kappa}} \hbar\omega_{\xi\mathbf{\kappa}} a^\dagger_{\xi\mathbf{\kappa}} a_{\xi\mathbf{\kappa}}$$

$$+ \sum_{\xi,\nu,\mathbf{K}} [\delta(\mathbf{\kappa} - \mathbf{K}) g_\nu(\xi\mathbf{\kappa}) b^\dagger_{\nu\mathbf{K}} a_{\xi\mathbf{\kappa}} + \text{h.c.}] \quad (9.36)$$

in which the first term is the Hamiltonian operator for the pure exciton field (Section 7) and the second is that of the pure photon field. The operators $a_{\xi\mathbf{\kappa}}$ and $a^\dagger_{\xi\mathbf{\kappa}}$ respectively annihilate and create photons of type $\xi\mathbf{\kappa}$. Their explicit effects on the functionals of the system are[234]

$$a_{\xi\mathbf{\kappa}} |\cdots N_{\nu\mathbf{K}}\cdots, \cdots n_{\xi\mathbf{\kappa}}\cdots\rangle = n^{1/2}_{\xi\mathbf{\kappa}} |\cdots N_{\nu\mathbf{K}}\cdots, \cdots n_{\xi\mathbf{\kappa}} - 1 \cdots\rangle \quad (9.37)$$

---

[232] D. L. Dexter, *Nuovo Cimento*, Suppl. **7**, 245 (1958).

[233] J. J. Hopfield, *Phys. Rev.* **112**, 1555 (1958).

[234] For explicit details, see, e.g., L. I. Schiff, "Quantum Mechanics," 2nd ed., Sect. 48. McGraw-Hill, New York, 1955.

and

$$a^\dagger_{\xi\kappa} | \cdots N_{\nu\mathbf{K}} \cdots, \cdots n_{\xi\kappa} \cdots \rangle = (n_{\xi\kappa} + 1)^{1/2} | \cdots N_{\nu\mathbf{K}} \cdots, \cdots n_{\xi\kappa} + 1 \cdots \rangle \quad (9.38)$$

with a similar convention for $b_{\nu\mathbf{K}}$ and $b^\dagger_{\nu\mathbf{K}}$, as introduced in Section 7. The coupling coefficient g can be computed as follows. Since the interaction term in Eq. (9.36) has the matrix element

$$n^{1/2}_{\xi\kappa} g_\nu(\xi\kappa)\, \bar{\delta}(\varkappa - \mathbf{K}) \quad (9.39)$$

between states $| \cdots 1_{\nu\mathbf{K}} \cdots, \cdots n_{\xi\kappa} - 1 \cdots \rangle$ and $| \cdots 0_{\nu\mathbf{K}} \cdots, \cdots n_{\xi\kappa} \cdots \rangle$ we see by comparison with Eq. (9.7), which is the corresponding matrix element in ordinary language, that

$$n^{1/2}_{\xi\kappa} g_\nu(\xi\varkappa) = (-e/mc)\, N^{1/2} A_0(\xi, \varkappa)\, \mathbf{e}_{\xi\kappa} \cdot (e^{i\boldsymbol{\kappa}\cdot\mathbf{r}} \mathbf{p})_{\nu\mathbf{K},0}\,. \quad (9.40)$$

Now $n(2\pi^2 c)^{-1} \omega^2_{\xi\kappa} | A_0 |^2$ is that part of the macroscopic energy density of the electromagnetic field at frequency $\omega_{\xi\kappa}$ which, in second quantization, is given by $n_{\xi\kappa} \hbar \omega_{\xi\kappa} V^{-1}$. It follows that

$$g_\nu(\xi\varkappa) = \frac{-e}{mc} \left(\frac{2\pi \hbar c}{\Omega_0 \kappa} \right)^{1/2} \mathbf{e}_{\xi\kappa} \cdot (e^{i\boldsymbol{\kappa}\cdot\mathbf{r}} \mathbf{p})_{\nu 0}\,, \quad (9.41)$$

where we have used $\omega = c\kappa/n$ to eliminate ω, and have put $V/N = \Omega_0$.

This linear interaction may be used in the formal theory of radiation to compute emission as well as absorption probabilities, but radiative emission is only one of the many channels by which excitons may decay. The linear interaction is a relatively strong one; it is of the order of $f^{1/2}(e^2/d)$, where d is a typical interatomic distance and f is the oscillator strength of the simple transition (9.1). For strong transitions, this interaction is of the order of magnitude of the exciton (or photon) energy itself. Fano and Hopfield eliminate the linear term by a unitary transformation on the zero-order states of the system; the result, given here in somewhat oversimplified terms, is that the eigenstates of the system with wave vector $\varkappa$ are best described as excitations of the form

$$A^\dagger_\kappa | \cdots 0 \cdots, \cdots 0 \cdots \rangle \quad (9.42)$$

where

$$A^\dagger_\kappa = c_1(\xi\varkappa)\, a^\dagger_{\xi\kappa} + c_2(\xi\varkappa)\, b^\dagger_{\nu\mathbf{K}} \quad (9.43)$$

creates a particle ("polariton") which is an intimate mixture of what we normally call photons and excitons, separately. A detailed derivation of c_1 and c_2 would carry us too far afield, but we note that they are determined mainly by the magnitude of g_ν and an "energy denominator," the difference between $hc\kappa$ and the usual exciton energy. Thus, excitons whose energies do not match those of a photon of the same wave vector are still good eigenstates of the system, as are photons whose energy and wave vector do not match those of an exciton. Since the dispersion curves of excitons and photons cross infrequently (Fig. 25), relatively few excitons are actually affected

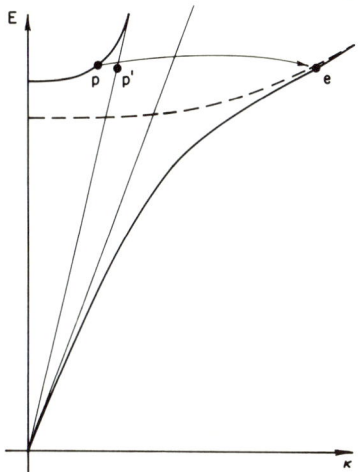

FIG. 25. Scattering of a "strongly mixed" particle p (polariton) into a relatively long-lived state e which is less likely to excite an external photon p'.

by the existence of polaritons. However, it is just these excitons which would be expected to be created by photons.

Most of the language of the Lorentz theory may now be converted into quantum-mechanical language: an external photon excites a crystal to a state described by a polariton packet, or superposition of states $A_\kappa^\dagger | 0 \rangle$. If this new particle of excitation fails to interact with any energy sinks, it re-excites an external photon on the other side of the crystal. However, by virtue of its exciton component, it can decay into states which do not couple as easily to external photons (i.e., "nonradiative" states). Such states might be reached by phonon and point defect scattering of the mixed particle. The presence of phonons or other imperfections is thus necessary for true absorption of light. This analysis does not invalidate the principal results of the semiclassical theory (Sections 9a, 9b); we can reinterpret them.[206] The probability per unit time that the system had made a transition into an exciton state (at the expense of a photon) was found to be proportional to the square of the exciton-photon coupling constant. This quantity now appears as a measure of the probability that a mixed-particle state will transfer its energy to all available sinks (long-lifetime states), in the sense that this total probability is proportional to the exciton amplitude in the polariton.

The fact that true crystal eigenstates are not pure excitons near $\mathbf{K} = 0$ has a bearing on the statistical arguments, as noted in Section 7a. If condensation takes place to the lowest energy state near $K = 0$, this would mean that the particles of excitation would "run down" the lower curve of Fig. 25, and would in a sense "condense out of the crystal," since the lower curve in this region represents mostly-photon-like states. It appears that the condensation must occur wholly within the upper branch of Fig. 25 or at some other point in the *BZ* where direct radiative annihilation is not a problem.

d. *Two-electron Transitions in Solids*[235]

An important process which is not predicted by a zero-order, one-electron description of a system[236] is a two-electron transition induced by a single photon. Long known in atomic spectra, particularly in the transition elements, this process has only recently been unambiguously observed in a solid,[237] where two electrons on *different* atoms are excited by a single photon. The origin of both of these transitions, in terms of the one-electron approximation, lies in first-order corrections to zero-order wave functions which are induced by the Coulomb interaction and can be called correlation effects. Normally one considers corrections to *energies* due to correlation effects; in atomic systems, we have "configuration interaction" and in molecular and solid systems, "van der Waals energies." These second-order corrections to energies have their first-order counterparts in the wave functions. Let $\psi_0^{(0)}$ and $\psi_{ab}^{(0)}$ be zero-order products of one-electron

[235] D. L. Dexter, *Phys. Rev.* **126**, 1962 (1962).
[236] See, e.g., E. U. Condon and G. H. Shortley, "The Theory of Atomic Spectra," Sect. 6^6, Chapters IX, XV. Cambridge Univ. Press, London and New York, 1953.
[237] F. Varsanyi and G. H. Dieke, *Phys. Rev. Letters* **7**, 442 (1961).

9. SEMICLASSICAL THEORY OF OPTICAL ABSORPTION

states corresponding, respectively, to the ground state (no excitations) and a two-electron excitation. Then the appropriate first-order states will contain *at least* the following terms:

$$\psi_0^{(1)} = \psi_0^{(0)} + \sum_{a',b' \neq 0} \lambda_{0,a'b'} \psi_{a'b'}^{(0)} + \cdots \quad (9.44)$$

and

$$\psi_{ab}^{(1)} = \psi_{ab}^{(0)} + \lambda_{ab,0} \psi_0^{(0)} + \cdots . \quad (9.45)$$

Here the λ are mixing coefficients, which consist of matrix elements of the Coulomb interaction between the states indicated, divided by appropriate energy denominators. The matrix element of the electric dipole operator between $\psi_0^{(0)}$ and $\psi_{ab}^{(0)}$ vanishes because this operator is a sum of one-electron operators.[235,236] However, the corresponding matrix element between the above first-order states does not necessarily vanish; consider, for example, the term of Eq. (9.44) in which $a' = a$. The state $\psi_{ab'}^{(0)}$ can be connected with $\psi_{ab}^{(0)}$ by a one-electron operator if the symmetries of excited states b and b' are appropriate. A full discussion of the various possible transitions which might occur in the case of single-photon excitation of two different atoms in a solid has been given by Dexter.[235]

The production of two excitons by a single photon can occur in precisely the same way. Ovander,[238] using some of the formalism of Agranovich,[206] concludes that the probability of creating two excitons is roughly a factor $(a/d)^3 f_{bb'}$ smaller than that for an allowed (single exciton) transition; where $f_{bb'}$ is the oscillator strength of a transition between the two states corresponding to $\psi_{ab}^{(0)}$ and $\psi_{ab'}^{(0)}$ in the above discussion, a is the "atomic orbit" size, and d the interatomic spacing. He feels that certain data on solid oxygen[239] is indicative of production of two excitons by one photon. Miyakawa[240] has extended Toyozawa's formalism[163] in a similar fashion in order to explain a peak in LiF observed by Milgram and Givens[110] at about 25 ev. One hopes that these exciton calculations will be refined in the near future to the point where definitive comparisons with experiment can be made; the exciton case is much harder to attack experimentally than that of

---

[238] L. N. Ovander, *Fiz. Tverd. Tela* **4**, 294 (1962); see *Soviet Phys.—Solid State (English Transl.)* **4**, 212 (1962).
[239] V. I. Dianov-Klokov, *Opt. i Spektroskopiya* **7**, 621 (1959).
[240] T. Miyakawa, *J. Phys. Soc. Japan* **17**, 1898 (1962).

two impurity atoms, since there is a vast number of other excitations possible in the region near twice the fundamental absorption energy.

An inverse process, one involving absorption of two photons to create one exciton, has been considered by Loudon.[241] The author is not aware of any direct tests of this mechanism as yet.

10. Dynamical Effects of Phonons

Except for the special case treated in Section 7, explicit introduction of any phonon effects has been avoided thus far. The limitations of the theory of Section 9 are due chiefly to disregard of the mechanism by which true absorption takes place; that theory provided for (at most) the *integrated* absorption coefficient associated with transitions to an isolated exciton state. In a vibrating but otherwise perfect crystal,

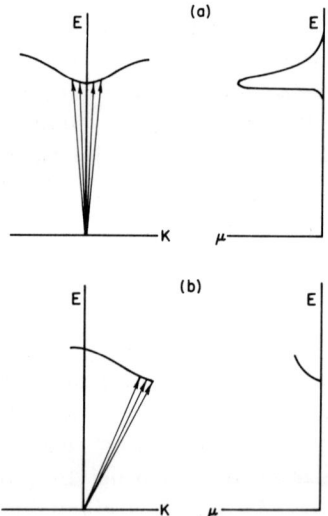

Fig. 26. Illustrating the effects of breakdown of the momentum selection rule on an exciton absorption spectrum. In (a), an original delta-function absorption is broadened. In (b), absorption appears where none is allowed at all in a static crystal. In the figure, E is the energy of the exciton state (left) or the corresponding photon (right); K is the usual wave vector, and μ an absorption coefficient in arbitrary units.

[241] R. Loudon, *Proc. Phys. Soc. (London)* **80**, 952 (1962). This work generalizes to excitons the one-electron theory of R. Braunstein, *Phys. Rev.* **125**, 475 (1962).

phonons are the natural choice of scatterers and sinks for the energy found in an exciton state. They broaden and shift the lines predicted by the elementary theory. The main reason for this is the fact that phonons relax the momentum selection rule and cause states in the neighborhood of $\mathbf{K} = \varkappa$ to be allowed to varying degrees. Relaxation of the momentum selection rule is also demonstrated explicitly by the existence of "indirect transitions," which are observed in cases in which the extremum of the exciton band is not located at $\mathbf{K} = 0$ (see Fig. 26). In Section 10a, the general exciton-phonon interaction is discussed. Sections 10b and 10c deal with line broadening and shifting, and Section 10d deals with indirect transitions.

a. The Exciton-Phonon Interaction[3,27,242–245]

In Section 2 it was assumed that the cores were at rest at the sites in a perfect lattice, and the valence-electron Hamiltonian $\mathscr{H}_0$ was so constructed. Now suppose that the atoms suffer small displacements from these sites, the displacement of atom I being written $\mathbf{u}_I$. Now the total crystal Hamiltonian becomes

$$\mathscr{H} = \mathscr{H}_0 + \sum_I (\nabla_I \mathscr{H})_0 \cdot \mathbf{u}_I + \tfrac{1}{2} \sum_I \sum_J \mathbf{u}_I \cdot (\nabla_I \nabla_J \mathscr{H})_0 \cdot \mathbf{u}_J + \cdots + \tfrac{1}{2} \sum_I M_I \ddot{\mathbf{u}}_I^2, \tag{10.1}$$

a Taylor expansion in the displacements, in which a subscript "0" indicates evaluation of a quantity at equilibrium. The last term in Eq. (10.1) is the kinetic energy of the atoms. Using the harmonic approximation,[246] we now take the total wave function of the system to be a product of an electronic state computed from $\mathscr{H}_0$ and a vibrational state, consisting of phonons, computed from the kinetic energy term and a dynamical matrix given by the expectation value of $(\nabla_I \nabla_J \mathscr{H})_0$ in the electronic state under consideration. Equilibrium of the lattice is established by the vanishing of the expectation value of the second term of Eq. (10.1). If the electronic state does not include

---

[242] A. I. Ansel'm and Yu. A. Firsov, *Zh. Eksperim. i. Teor. Fiz.* **28**, 151 (1955); see *Soviet Phys. J.E.T.P. (English Transl.)* **1**, 139 (1955).

[243] A. I. Ansel'm and Yu. A. Firsov, *Zh. Eksperim. i Teor. Fiz.* **30**, 719 (1956); see *Soviet Phys. J.E.T.P. (English Transl.)* **3**, 564 (1956).

[244] Y. Toyozawa, *Progr. Theoret. Phys. (Kyoto), Suppl.* **12**, 111 (1959).

[245] Y. Toyozawa, *Progr. Theoret. Phys. (Kyoto)* **27**, 89 (1962).

[246] See, e.g., M. Born and K. Huang, "Dynamical Theory of Crystal Lattices," Section 14. Oxford Univ. Press, London and New York, 1954.

too many excitons, the dynamical matrix is essentially determined by the electronic charge distribution of the ground state. Thus there is a unique phonon spectrum, independent of the excitons; this is true only in the case of ordinary excitons, i.e., those which have not become trapped at definite lattice sites. If trapping occurs (Section 12), the resulting local distortion requires a change in this normal mode description.

Returning to Eq. (10.1), we now make use of the (phonon) normal coordinate transformation

$$q_{\sigma \mathbf{f}} = N^{-1/2} \sum_i \sum_I c_{\sigma i}(\mathbf{f}) \, e^{i\mathbf{f} \cdot \mathbf{R}_I} M_{Ii}^{1/2} u_{Ii} \qquad (10.2)$$

where $\mathbf{f}$ is the phonon wave vector, σ labels its polarization, and c is a set of coupling parameters known in principle. Here, I is to be thought of as a unit cell index, and i runs over all the Cartesian coordinates of all the atoms within the unit cell. The normal coordinate $q_{\sigma \mathbf{f}}$ corresponds to a normal mode frequency $\omega_{\sigma \mathbf{f}}$, so the total Hamiltonian now becomes

$$\mathscr{H} = \mathscr{H}_0 + \sum_{\sigma, \mathbf{f}} (\partial \mathscr{H}/\partial q_{\sigma \mathbf{f}})_0 \, q_{\sigma \mathbf{f}} + \tfrac{1}{2} \sum_{\sigma, \mathbf{f}} (\dot{q}^\dagger_{\sigma, \mathbf{f}} \dot{q}_{\sigma, \mathbf{f}} + \omega^2_{\sigma \mathbf{f}} q^\dagger_{\sigma, \mathbf{f}} q_{\sigma, \mathbf{f}}). \qquad (10.3)$$

Anharmonic terms have been neglected, and the simplicity of the second term of Eq. (10.3) is a result of the fact that the form $\Sigma_I \mathbf{u}_I \cdot \nabla_I$ is invariant under a unitary transformation like Eq. (10.2). After quantization of the vibrational Hamiltonian, the first and third terms of Eq. (10.3) will be diagonal in all electronic and phonon quantum numbers; the second term will provide the exciton-phonon interaction. This may be seen by noting that $(\partial \mathscr{H}/\partial q_{\sigma, \mathbf{f}})_0$ does not have the full translational symmetry of the lattice. It acquires a phase factor $e^{-i\mathbf{f} \cdot \mathbf{m}}$ under a lattice translation $\mathbf{m}$, and therefore couples electronic states (and, in particular, excitons) whose total wave vectors differ by $\mathbf{f}$ plus 2π times a reciprocal lattice vector. This will be seen explicitly in our subsequent calculation. We shall first define an exciton-phonon coupling constant G in terms of the matrix element of $(\partial \mathscr{H}/\partial q_{\sigma \mathbf{f}})_0$ between two general exciton states:

$$\int \Psi^*_{\nu \mathbf{K}} (\partial \mathscr{H}/\partial q_{\sigma \mathbf{f}})_0 \, \Psi_{\nu' \mathbf{K}'} \, d\tau \equiv \bar{\delta}(\mathbf{f} - \mathbf{K} + \mathbf{K}') (2\omega_{\sigma \mathbf{f}}/\hbar)^{1/2} \, G_{\sigma \mathbf{f}}(\nu \mathbf{K}, \nu' \mathbf{K}'). \qquad (10.4)$$

Since this matrix element deals with the scattering of an exciton from a state $\nu'\mathbf{K}'$ into a state $\nu\mathbf{K}$, we may write it in a second-quantized notation as a sum of all possible processes of this type,

$$\sum_{\nu\mathbf{K}}\sum_{\nu'\mathbf{K}'}\bar{\delta}(\mathbf{f}-\mathbf{K}+\mathbf{K}')\,(2\omega_{\sigma\mathbf{f}}/\hbar)^{1/2}\,G_{\sigma\mathbf{f}}(\nu\mathbf{K},\nu'\mathbf{K}')\,b^{\dagger}_{\nu\mathbf{K}}b_{\nu'\mathbf{K}'}. \qquad (10.5)$$

It is quite convenient to proceed to a completely second-quantized formulation by introducing phonon creation and annihilation operators $\alpha^{\dagger}_{\sigma\mathbf{f}}$ and $\alpha_{\sigma\mathbf{f}}$:

$$q_{\sigma\mathbf{f}} = i(\hbar/2\omega_{\sigma\mathbf{f}})^{1/2}\,(\alpha_{\sigma\mathbf{f}} - \alpha^{\dagger}_{\sigma,-\mathbf{f}}) \qquad (10.6)$$

$$\dot{q}_{\sigma\mathbf{f}} = (\hbar\omega_{\sigma\mathbf{f}}/2)^{1/2}\,(\alpha_{\sigma\mathbf{f}} + \alpha^{\dagger}_{\sigma,-\mathbf{f}}). \qquad (10.7)$$

Since they are ordinary boson operators, $\alpha_{\sigma\mathbf{f}}$ and $\alpha^{\dagger}_{\sigma\mathbf{f}}$ have the effect of decreasing and increasing by one the numbers of phonons in the mode $\sigma\mathbf{f}$ and multiplying the system functional by $n_{\sigma\mathbf{f}}^{1/2}$ and $(n_{\sigma\mathbf{f}}+1)^{1/2}$, respectively, where $n_{\sigma\mathbf{f}}$ is the number of phonons originally in the mode $\sigma\mathbf{f}$. Now Eq. (10.3) becomes

$$\mathcal{H} = \sum_{\nu\mathbf{K}} E_{\nu\mathbf{K}} b^{\dagger}_{\nu\mathbf{K}} b_{\nu\mathbf{K}} + \sum_{\sigma\mathbf{f}} \hbar\omega_{\sigma\mathbf{f}}(\alpha^{\dagger}_{\sigma\mathbf{f}}\alpha_{\sigma\mathbf{f}} + \tfrac{1}{2})$$

$$+ \sum_{\nu\mathbf{K}}\sum_{\nu'\mathbf{K}'}\sum_{\sigma\mathbf{f}}\bar{\delta}(\mathbf{f}-\mathbf{K}+\mathbf{K}')\,iG_{\sigma\mathbf{f}}(\nu\mathbf{K},\nu'\mathbf{K}')\,b^{\dagger}_{\nu\mathbf{K}}b_{\nu'\mathbf{K}'}(\alpha_{\sigma\mathbf{f}} - \alpha^{\dagger}_{\sigma,-\mathbf{f}}). \qquad (10.8)$$

The exciton-lattice Hamiltonian has been used in this (or an equivalent) form by several authors, and we will regard it as the standard form. It should be emphasized that the interaction term is linear in phonon operators and this Hamiltonian may be inadequate for extremely strong coupling.

The term in Eq. (10.8) involving $\alpha^{\dagger}_{\sigma,-\mathbf{f}}$ creates a phonon of wave vector $-\mathbf{f} = \mathbf{K}' - \mathbf{K}$, and scatters an exciton from $\mathbf{K}'$ to $\mathbf{K}$; the term involving $\alpha_{\sigma\mathbf{f}}$ destroys a phonon of wave vector $\mathbf{f} = \mathbf{K} - \mathbf{K}'$ and likewise scatters an exciton from $\mathbf{K}'$ to $\mathbf{K}$. Energy conservation does not appear explicitly in the Hamiltonian, because we are using the Schrödinger picture and the time dependence is contained in the state vectors rather than in the operators. Because of energy conservation, too, we do not include terms linear in exciton operators; generally, the most energetic phonon available has much less energy than the lowest possible exciton.

The coupling "constant" G depends markedly on the type of phonon involved, particularly on whether it is acoustical or optical;

it also depends on the details of the exciton wave function. It has been worked out under various simplifying assumptions, as follows.

(1) *Frenkel Case.* By direct substitution of the Frenkel wave function (3.5) into Eq. (10.4), and using the gradient operator

$$\frac{\partial}{\partial q_{0\mathbf{f}}} = \sum_i c_{\sigma i}(\mathbf{f})^* (M_i N)^{-1/2} \sum_I e^{-i\mathbf{f}\cdot\mathbf{R}_I} \frac{\partial}{\partial u_{Ii}} \qquad (10.9)$$

we obtain, neglecting the possibility of Umklapp processes,

$$G_{\sigma,\mathbf{K}'-\mathbf{K}}(\nu\mathbf{K}, \nu'\mathbf{K}') = \sum_i c_{\sigma i}(\mathbf{K}' - \mathbf{K}) (\hbar/2\omega_{\sigma,\mathbf{K}'-\mathbf{K}} M_i N)^{1/2}$$

$$\times \sum_{\mathbf{R}} \sum_{\mathbf{R}'} e^{-i(\mathbf{K}\cdot\mathbf{R}-\mathbf{K}'\cdot\mathbf{R}')} \int \Phi_\nu(\mathbf{R})^* (\partial\mathcal{H}/\partial u_{0i})_0 \Phi_{\nu'}(\mathbf{R}') \, d\tau. \qquad (10.10)$$

The abbreviated notation of Section 3 has been used, i.e., $\Phi_\nu(\mathbf{R})$ is a localized excited state in the cell at $\mathbf{R}$. In reasonably simple models, only terms involving $\mathbf{R} = 0$ and $\mathbf{R}' = 0$ will be large; other terms immediately bring three-center integrals into the calculation. The term with $\mathbf{R} = \mathbf{R}' = 0$ may sometimes be important but appears to vanish in most cases by symmetry when $\nu = \nu'$. The terms remaining in Eq. (10.10) are then

$$\sum_i c_{\sigma i}(\mathbf{K}' - \mathbf{K}) (\hbar/2\omega_{\sigma,\mathbf{K}'-\mathbf{K}} M_i N)^{1/2} \left[\sum_\mathbf{R} e^{-i\mathbf{K}\cdot\mathbf{R}} \int \Phi_\nu(\mathbf{R})^* (\partial\mathcal{H}/\partial u_{0i})_0 \Phi_{\nu'}(0) \, d\tau \right.$$

$$\left. + \sum_{\mathbf{R}'\neq 0} e^{i\mathbf{K}'\cdot\mathbf{R}'} \int \Phi_\nu(0)^* (\partial\mathcal{H}/\partial u_{0i})_0 \Phi_{\nu'}(\mathbf{R}') \, d\tau \right]. \qquad (10.11)$$

The integrals in square brackets reduce, in the dipole-dipole approximation (Section 3), to gradients of the dipole-dipole interaction which produces the Frenkel-exciton band structure. Agranovich and Konobeev[247] have computed the term in square brackets for the case of a simple cubic array of lattice constant d with excited states whose dipoles lie along the z axis, with oscillator strength $f_{\nu 0}$ at energy $\hbar\omega_{\nu 0}$.

[247] V. M. Agranovich and Yu. V. Konobeev, *Opt. i Spektroskopiya* **6**, 242 (1959).

10. DYNAMICAL EFFECTS OF PHONONS

Taking only the nearest neighbors in the sums, they obtain

$$G_{\sigma,\mathbf{K}'-\mathbf{K}}(\nu\mathbf{K}, \nu'\mathbf{K}') = \sum_j c_{\sigma j}(\mathbf{K}' - \mathbf{K})\left(\frac{\hbar}{2\omega_{\sigma,\mathbf{K}'-\mathbf{K}} M_j N}\right)^{1/2}$$

$$\times \left\{i\frac{3e^2 f_{\nu 0}\hbar}{m\omega_{\nu 0}d^3}\left[(\mathbf{K}' - \mathbf{K})_j - 3(\mathbf{K}' - \mathbf{K})_j \delta_{jz}\right]\right\}. \quad (10.12)$$

Here j runs over only the coordinates x, y, z of the one atom in the unit cell; only longitudinal phonons interact with the exciton, since the summand depends only on the j component of $\mathbf{K}' - \mathbf{K} = \mathbf{f}$. The sums in Eq. (10.11) can be carried out in the continuum approximation,[39] and one obtains a fairly complicated expression which does *not* depend on $\mathbf{K}$ and $\mathbf{K}'$ only through their difference. Thus transverse phonons may well contribute to the interaction. Some caution must be exercised in using Eq. (10.12) in isotropic or cubic crystals; it connects states whose transition dipoles are *parallel*, and therefore may be useful only in indirect transitions where one state involved is an intermediate state. It is in general not a matrix element between two *good* eigenstates. Further details on the Frenkel coupling constant may be found in the work of Davydov,[248] Haga and Yokota,[249] and Simpson and Peterson.[250]

(2) *Wannier Case.* Here a slightly different attack is made which emphasizes the nearly independent behavior of the electron and hole as effective-mass particles. Lattice vibrations are specifically assumed to affect $\mathscr{H}$ only through single-particle interactions; thus we assume that the interaction term in Eq. (10.3) may be written

$$\sum_{\sigma f} (\partial \mathscr{H}/\partial q_{\sigma f}) q_{\sigma f} = \sum_i \delta \mathscr{H}(\mathbf{r}_i). \quad (10.13)$$

Using the general exciton states (2.31), the matrix element of this interaction between two exciton states becomes

$$\int \Psi_{\nu\mathbf{K}}^* \sum_{\sigma f}(\partial \mathscr{H}/\partial q_{\sigma f})_0 q_{\sigma f} \Psi_{\nu'\mathbf{K}'} d\tau$$

$$= \sum_{\beta}\sum_{\beta'} U_{mn\nu\mathbf{K}}(\beta)^* U_{m'n'\nu'\mathbf{K}'}(\beta') \int \Phi_{mn}(\mathbf{K}\beta)^* \left[\sum_i \delta\mathscr{H}(\mathbf{r}_i)\right] \Phi_{m'n'}(\mathbf{K}'\beta') d\tau.$$

(10.14)

[248] A. S. Davydov, *Tr. Inst. Fiz., Akad. Nauk Ukr. S.S.R.* **3**, 36 (1952); see *Naval Res. Lab. Transl.* No. 867 (1961).
[249] E. Haga and I. Yokota, *Busseiron-Kenkyu* **90**, 50 (1955).
[250] W. T. Simpson and D. L. Peterson, *J. Chem. Phys.* **26**, 588 (1957).

We now use the connection between the exciton and Bloch representation, Eq. (2.27), to express (10.14) in terms of one-electron matrix elements of $\delta \mathcal{H}(\mathbf{r})$. The result is

$$\sum_{\beta} \sum_{\beta'} U_{mn\nu\mathbf{K}}(\beta)^* U_{m'n'\nu'\mathbf{K}'}(\beta') \left[-N^{-1} \sum_{\mathbf{k}} e^{i(\beta-\beta')\cdot\mathbf{k}} \, \Xi_{mm'}(\mathbf{k}-\mathbf{K}',\mathbf{k}-\mathbf{K}) \right.$$

$$\left. + N^{-1} \sum_{\mathbf{k}} e^{i(\beta-\beta')\cdot\mathbf{k}+i\beta'(\mathbf{K}-\mathbf{K}')} \, \Xi_{nn'}(\mathbf{k},\mathbf{k}-\mathbf{K}+\mathbf{K}') \right] \quad (10.15)$$

where

$$\Xi_{mm'}(\mathbf{k},\mathbf{k}') = \int \psi_{m\mathbf{k}}(\mathbf{r})^* \, \delta \mathcal{H}(\mathbf{r}) \, \psi_{m'\mathbf{k}'}(\mathbf{r}) \, d\tau \quad (10.16)$$

is the ordinary scattering matrix element between Bloch functions. Thus we may draw directly on the literature of band theory in obtaining coupling constants. Deformation potential theory[251] yields

$$\Xi_{mm'}(\mathbf{k},\mathbf{k}') = \sum_{\sigma\mathbf{f}} \sum_{j} c_{\sigma j}(\mathbf{f})^* \, (\hbar/2M_j N\omega_{\sigma\mathbf{f}})^{1/2}$$

$$\times \delta(\mathbf{f}-\mathbf{k}+\mathbf{k}') \, i^{-1} f_j E_{mm'}(\alpha_{\sigma\mathbf{f}} - \alpha^\dagger_{\sigma,-\mathbf{f}}) \quad (10.17)$$

in the case of acoustic phonons; $E_{mm'}$ is the matrix element of the deformation potential between bands m and m', and f_j is the j component of $\mathbf{f}$. If optical phonons in polar crystals are involved, the Frohlich interaction[252] may be used. Here

$$\Xi_{mm'}(\mathbf{k},\mathbf{k}') = \sum_{\sigma\mathbf{f}} \sum_{j} c_{\sigma j}(\mathbf{f})^* \, \frac{1}{i} \left[\frac{2\hbar\omega_{\sigma\mathbf{f}}\pi}{N\Omega_0} \right]^{1/2}$$

$$\times \left(\frac{1}{\epsilon_\infty} - \frac{1}{\epsilon_0} \right)^{1/2} \frac{(\mathbf{k}-\mathbf{k}')_j}{|\mathbf{k}-\mathbf{k}'|^2} \, \delta(\mathbf{f}-\mathbf{k}-\mathbf{k}') (\alpha_{\sigma\mathbf{f}} - \alpha^\dagger_{\sigma,-\mathbf{f}}). \quad (10.18)$$

In this case, the index σ runs only over the optical modes and Σ_j selects the longitudinal mode, as in the acoustic case. Here ϵ_∞, ϵ_0, and Ω_0 are the high-frequency and low-frequency dielectric constants and the unit cell volume, respectively. It is of course, worthwhile to note that this is the interaction between bare charged

---

[251] J. Bardeen and W. Shockley, *Phys. Rev.* **80**, 72 (1950).
[252] See, e.g., H. Frohlich et al.[170]

particles and phonons; it is necessary to "clothe" the electron and hole if a completely quantitative theory is desired.[253]

An important feature of the foregoing interactions is the fact that Ξ depends only on the difference $\mathbf{k} - \mathbf{k}'$, since it leads to considerable simplification of the exciton calculation. We write

$$\Xi(\mathbf{k}, \mathbf{k}') = \Xi(\mathbf{k} - \mathbf{k}')$$

and obtain for Eq. (10.15)

$$\int \Psi^*_{\nu\mathbf{K}} \sum_{\sigma f} (\partial\mathcal{H}/\partial q_{\sigma f})_0 \, q_{\sigma f} \, \Psi_{\nu'\mathbf{K}'} \, d\tau$$
$$= - q_h(mn\nu\mathbf{K}, m'n'\nu'\mathbf{K}') \, \Xi_{mm'}(\mathbf{K} - \mathbf{K}') + q_e(mn\nu\mathbf{K}, m'n'\nu'\mathbf{K}') \, \Xi_{nn'}(\mathbf{K} - \mathbf{K}') \quad (10.19)$$

where q_h and q_e are factors of order zero to unity, depending on the extent of the overlap of the Wannier envelope functions involved:

$$q_h = \sum_{\beta} U_{mn\nu\mathbf{K}}(\beta)^* \, U_{m'n'\nu'\mathbf{K}'}(\beta) \quad (10.20)$$

and

$$q_e = \sum_{\beta} U_{mn\nu\mathbf{K}}(\beta)^* \, U_{m'n'\nu'\mathbf{K}'}(\beta) \, e^{i\beta\cdot(\mathbf{K}-\mathbf{K}')}. \quad (10.21)$$

These factors, apparently first introduced by Ansel'm and Firsov in a different notation,[242,243] are functions only of $\mathbf{K} - \mathbf{K}'$ in the effective mass approximation when $mn = m'n'$, regardless of the positions of the electron and hole extrema in the bands, and represent the Fourier transforms of the hydrogenic envelope function overlap distributions $F_\nu(\beta)^* \, F_{\nu'}(\beta)$. Toyozawa discusses them as measures of the "effectiveness" of the electron and hole in the exciton for scattering by a phonon of wave number $\mathbf{K} - \mathbf{K}'$.

It is now possible to combine Eq. (10.19) with (10.4), using the specific one-electron interactions (10.17) and (10.18), in order to obtain $G_{\sigma f}$. In the case of scattering between excitons built from the same pair of conduction and valence bands ($n = n'$ and $m = m'$), the exciton-acoustic mode coupling constant is

$$G_{ac,\mathbf{K}'-\mathbf{K}}(\nu\mathbf{K}, \nu'\mathbf{K}') = (\hbar/2MN u)^{1/2} \, | \, \mathbf{K}' - \mathbf{K} \, |^{1/2}$$
$$\times [q_h(mn\nu\mathbf{K}, mn\nu'\mathbf{K}') \, E_{mm} - q_e(mn\nu\mathbf{K}, mn\nu'\mathbf{K}') \, E_{nn}] \quad (10.22)$$

---

[253] See, e.g., T. D. Schultz, *Phys. Rev.* **116**, 526 (1959); several references to reviews of polaron physics are found in this paper.

where M is the total mass of the unit cell, u is the velocity of sound, and E_{ll} is the deformation potential for the lth band. We have used the approximation $\omega_{ac,\mathbf{f}} = uf$, valid for $\mathbf{f}$ near the center of the BZ. Finally, in the case of the exciton-optical phonon interaction, we have

$$G_{op,\mathbf{K'}-\mathbf{K}}(\nu\mathbf{K}, \nu'\mathbf{K'}) = \left(\frac{2\pi\hbar\omega_0 e^2}{N\Omega_0}\right)\left(\frac{1}{\epsilon_\infty} - \frac{1}{\epsilon_0}\right)^{1/2}$$

$$\times \frac{1}{|\mathbf{K}-\mathbf{K'}|}[-q_h(mn\nu\mathbf{K}, mn\nu'\mathbf{K'}) + q_e(mn\nu\mathbf{K}, mn\nu'\mathbf{K'})] \quad (10.23)$$

where ω_0 is the longitudinal optical phonon frequency at $\mathbf{f} \sim 0$.

We shall discuss the orders of magnitude and other properties of these coupling parameters as the occasion arises.

b. Line Broadening

Immediately after the appearance of Frenkel's original papers, Peierls[3] published a study of the effects of phonons with a linear interaction (10.8), classifying solids into "scatterers" and "absorbers," according to whether the linear approximation did or did not appear to be valid. Scatterers, Peierls showed, could not turn photons directly into lattice vibrations, because too high an order of perturbation theory (too many phonons) would be required to conserve energy; an "absorption spectrum" would occur, but all the photons would be reemitted at essentially their original energies. Frenkel reinterpreted[4] Peierls's "absorbing" case in terms of *trapping* of the excitons, a manifestation of very strong exciton-lattice coupling which goes so far beyond the applicability of the linear approximation that new equilibrium positions must be considered for the nuclei when the electronic system is excited. He speculated that scatterers in Peierls's sense would not exist, and nothing resembling them were in fact discovered until quite recently; we discuss these cases in Section 13. Frenkel concluded that in most cases a photon would create a wave packet of excitation which would scatter frequently, becoming trapped before emerging from the crystal.

(1) *Perturbation Approach*.[242,243,247,250,254,255] The intricacies of computing an absorption coefficient as a function of incident photon

---

[254] P. J. Leurgans and J. Bardeen, *Phys. Rev.* **87**, 200 (1952); P. J. Leurgans, Thesis, University of Illinois, 1952.

[255] V. I. Davydov and B. Shmushkevich, *Zh. Eksperim. i Teor. Fiz.* **10**, 1043 (1940).

energy have been nicely detailed by Toyozawa.[244] One must couple a simple photon field on the outside of the crystal to a highly complicated field on the inside, the latter field being governed by the total Hamiltonian

$$\mathcal{H} = \mathcal{H}_0 + \mathcal{H}_{eR} + \mathcal{H}_R + \mathcal{H}_{eL} + \mathcal{H}_L \tag{10.24}$$

where $\mathcal{H}_0$ is the electronic Hamiltonian, $\mathcal{H}_R$ is that of the radiation field, $\mathcal{H}_L$ is that of the lattice, $\mathcal{H}_{eR}$ is the electron-radiation coupling [the third term of Eq. (9.36)] and $\mathcal{H}_{eL}$ is the electron-lattice coupling derived in Section 10a. One can treat either $\mathcal{H}_{eR}$ or $\mathcal{H}_{eL}$ as a weak perturbation, but the results obtained are of fairly restricted applicability. We know, for example (Section 9c) that (when oscillator strengths are large) the first 3 terms of Eq. (10.24) must be considered together to obtain an unobjectionable description of the excitations of the crystal in the absence of $\mathcal{H}_{eL}$. Despite this, a treatment assuming that both $\mathcal{H}_{eL}$ and $\mathcal{H}_{eR}$ are small perturbations is instructive and yields much qualitative information about absorption shapes. We therefore consider states of the form | photons; excitons; phonons⟩ and compute by semiclassical radiation theory the probability of transitions of the type

$$| 1_{\xi\kappa}; 0; \cdots n_{\sigma f} \cdots \rangle \rightarrow | 0; 1_{\nu K}; \cdots n'_{\sigma f} \cdots \rangle. \tag{10.25}$$

(Here, and throughout this section, $\varkappa \approx 0$ is the wave vector of the perturbing light.) This is just a transition of the type (9.1) with the added possibility that the phonon population may change. If the linear exciton-phonon interaction in Eq. (10.8) is used, this change will amount to only the creation (or annihilation) of one phonon, and by conservation of momentum its wave vector will be $\varkappa - \mathbf{K}$ or $-(\varkappa - \mathbf{K})$, neglecting umklapp processes. We compute first-order wave functions using $\mathcal{H}_{eL}$, which affects only $| 0; 1_{\nu K}; \cdots n'_{\sigma f} \cdots \rangle$, and then substitute them into the expression for the transition probability

$$w = (2\pi/\hbar)\, Av \sum | \langle 1_{\xi\kappa}; 0; \cdots n_{\sigma f} \cdots | \mathcal{H}_{eR} | 0; 1_{\nu K}; \cdots n'_{\sigma f} \cdots \rangle |^2$$
$$\times \delta(\hbar\omega - E_{\nu K} \pm \hbar\omega_{\sigma f}). \tag{10.26}$$

Here the average is a thermal average and is taken over the initial states and the sum is taken over final states. In addition to a term involving no changes in phonons, which was the subject of Section 9, we obtain

$$\frac{2\pi}{\hbar} \sum_{\sigma \mathbf{K}} \left| \sum_{\nu'} g_{\nu'}(\xi \varkappa) \frac{G_{\sigma, \mathbf{K}-\varkappa}(\nu'\varkappa, \nu \mathbf{K})}{E_{\nu \mathbf{K}} - E_{\nu' \varkappa} + \hbar\omega_{\sigma, \mathbf{K}-\varkappa}} \right|^2 (\bar{n}_{\sigma, \mathbf{K}-\varkappa} + 1)$$

$$\times \delta[\hbar\omega - (E_{\nu \mathbf{K}} + \hbar\omega_{\sigma, \mathbf{K}-\varkappa})]$$

$$+ \frac{2\pi}{\hbar} \sum_{\sigma \mathbf{K}} \left| \sum_{\nu'} g_{\nu'}(\xi \varkappa) \frac{G_{\sigma, \mathbf{K}-\varkappa}(\nu'\varkappa, \nu \mathbf{K})}{E_{\nu \mathbf{K}} - E_{\nu' \varkappa} - \hbar\omega_{\sigma, \mathbf{K}-\varkappa}} \right|^2 \bar{n}_{\sigma, \mathbf{K}-\varkappa}$$

$$\times \delta[\hbar\omega - (E_{\nu \mathbf{K}} - \hbar\omega_{\sigma, \mathbf{K}-\varkappa})] \quad (10.27)$$

where $\bar{n}_{\sigma f} = [\exp(\hbar\omega_{\sigma f}/kT) - 1]^{-1}$. At this stage of perturbation theory[256] the direct absorption line is unchanged, and according to Eq. (10.27) there appear in addition to it a large number of satellite lines at energies $E_{\nu \mathbf{K}} \pm \hbar\omega_{\sigma f}$. These lines will have their own natural widths and are nearly continuously distributed, so that a smooth absorption band results.

The terms in Eq. (10.27) with $\nu = \nu'$ may be called *intraband* scattering terms, while those with $\nu \neq \nu'$ are *interband* terms. Some of their roles are shown in Fig. 27. The first example, Fig. 27a, is

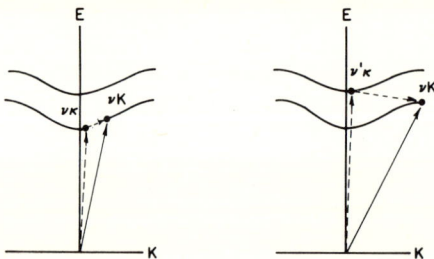

FIG. 27. (a) Intraband and (b) interband exciton scattering by phonons.

[256] We have used first-order time-dependent perturbation theory to reach states computed in first-order time-independent perturbation theory with a different perturbation. The calculation is thus actually a second-order calculation, and if it were carried out completely we would find that the direct absorption peak would have lost part of its strength to the satellites, such that the total strength of the band is equal to that given by Section 9 for the delta function absorption. This is a special case of an f-sum rule, first given by Peierls[3] for excitons and familiar in other fields [see J. H. Van Vleck, *Phys. Rev.* **74**, 1168 (1948); M. Lax, *J. Chem. Phys.* **20**, 1752 (1952)]. It is valid for allowed, isolated transitions.

simple line broadening, in which a state at energy $E_{\nu K} \pm \hbar\omega_{\sigma,\mathbf{K}-\mathbf{\kappa}}$ in the same band as the central line is reached with the help of a phonon of fairly small wave vector. The dotted line shows that $\mathscr{H}_{eR}$ connects the ground state to the exciton state responsible for the central line, $\Psi_{\nu\kappa}$, which acts as an intermediate state. As the energy moves farther from that of the central line, the energy denominators in Eq. (10.27) reduce the oscillator strengths of the satellites, forming the "tail" of the absorption band. In Fig. 27b, a phonon of rather large wave vector and an intermediate state $\Psi_{\nu'\mathbf{K}}$ combine to allow an "indirect" exciton transition to take place to a final state high in band ν.[257] This is conceivably an important process, now, because the energy denominator may again be small. It competes with the direct transition to $\Psi_{\nu'\mathbf{\kappa}}$ which is also taking place at this energy, and can contribute to broadening of the line associated with band ν'. It has been suggested[258] that processes of this type are responsible for the large breadth of the Cu_2O $2s$ line (compared with that of the $1s$ line).

The computation of absorption shapes using Eq. (10.27) is unwieldy, and weak-coupling perturbation theory is usually carried through in a slightly different manner. One merely regards the state $|0; 1_{\nu\kappa}; \cdots n_{\sigma f} \cdots\rangle$ as having a finite lifetime against transitions to other states $|0; 1_{\nu'\mathbf{K}'}; \cdots n_{\sigma f'} \cdots\rangle$, these transitions being induced by $\mathscr{H}_{eL}$. Then by ordinary first-order theory, the reciprocal of this lifetime is

$$\Gamma_{\nu\kappa} = (2\pi/\hbar) \sum_{\sigma\nu'\mathbf{K}'}' |G_{\sigma,\mathbf{K}'-\mathbf{\kappa}}(\nu'\mathbf{K}', \nu\kappa)|^2$$
$$\times [(\bar{n}_{\sigma,\mathbf{\kappa}-\mathbf{K}'} + 1) \delta(E_{\nu\kappa} - E_{\nu'\mathbf{K}'} - \hbar\omega_{\sigma,\mathbf{K}'-\mathbf{\kappa}})$$
$$+ \bar{n}_{\sigma,\mathbf{K}'-\mathbf{\kappa}} \delta(E_{\nu\kappa} - E_{\nu'\mathbf{K}'} + \hbar\omega_{\sigma,\mathbf{K}-\mathbf{K}'})]. \quad (10.28)$$

The line width[259] is then taken to be $\hbar\Gamma_{\nu\kappa}$. In addition, the shift of the state $\Psi_{\nu\kappa}$ can be computed by second-order stationary perturbation theory, with the result

$$\Delta_{\nu\kappa} = -\sum_{\sigma\nu'\mathbf{K}'}' |G_{\sigma,\mathbf{K}'-\mathbf{\kappa}}(\nu'\mathbf{K}', \nu\kappa)|^2$$
$$\times \left(\frac{\bar{n}_{\sigma,\mathbf{K}'-\mathbf{\kappa}}}{E_{\nu'\mathbf{K}'} - E_{\nu\kappa} - \hbar\omega_{\sigma,\mathbf{K}'-\mathbf{\kappa}}} + \frac{\bar{n}_{\sigma,\mathbf{K}'-\mathbf{\kappa}} + 1}{E_{\nu'\mathbf{K}'} - E_{\nu\kappa} + \hbar\omega_{\sigma,\mathbf{K}'-\mathbf{\kappa}}} \right). \quad (10.29)$$

[257] This example differs from the indirect processes of Section 10d mainly in that the bottom of band ν is not at the edge of the zone in the present case.

[258] R. J. Elliott, *Proc. Intern. Conf. Semicond. Phys., Prague, 1960,* p. 408 (1961).

[259] The validity of this procedure has been established in detail by R. Brout, *Phys. Rev.* **107**, 664 (1957).

In performing the summation over $\mathbf{K}'$, a principal value must be taken because of singularities in the summand.

Lifetimes of both Frenkel and Wannier excitons have been calculated using Eq. (10.28). We reserve the Frenkel case for a later section and quote here only some results for scattering by thermal acoustic phonons, where the integrations are not too complicated. For a Wannier 1s exciton having a wavelength larger than $\hbar/M^*u \sim 100$ A and scattering to excitons in its own band it is found that[27,242,243]

$$\Gamma_{\nu\mathbf{K}} = \frac{8(M^*)^2 \, kT(E_{mm} - E_{nn})^2 \, \Omega_0}{9\pi\hbar^4 M u}. \tag{10.30}$$

Here $M^* = m_e^* + m_h^*$ is the exciton effective mass, and the other symbols are defined in Section 10a. For excitons whose wavelengths are smaller than $\hbar/M^*u$ but still longer than the exciton radius, $\Gamma_{\nu\mathbf{K}}$ becomes $\mathbf{K}$-dependent and equal to

$$\Gamma_{\nu\mathbf{K}} = \frac{4M^* kT(E_{mm} - E_{nn})^2 \, \Omega_0}{9\pi\hbar^3 M u^2} \cdot \frac{|\mathbf{K} - \mathbf{K}_0|^2 + (M^*u/\hbar)^2}{|\mathbf{K} - \mathbf{K}_0|} \tag{10.31}$$

where $\mathbf{K}_0$ is the position of the minimum of the band. These values of $\Gamma_{\nu\mathbf{K}}$ are high-temperature results and are not linear all the way to $T = 0$. Low-temperature results are quoted by Toyozawa; Genkin[260] has done computations on higher Wannier states at high temperatures.

Detailed comparison of these results with experiment has not generally been made, mainly because of a lack of all the necessary data, and partly because the theory has not been pushed far enough to account for degeneracies and other complications in the band structures of crystals where the necessary parameters are known. Early calculations by Leurgans and Bardeen,[254] giving mean free paths of 100 to 3000 A in Ge, were made on an early standard-band picture of that solid. Toyozawa finds reasonable agreement between his results and data by Martienssen on alkali halides,[261] and Gold and the present author have attempted similarly to analyze data on the solid rare gases.[262] Only order-of-magnitude agreement is achieved. We may also note that Eqs. (10.30) and (10.31) lead,

[260] G. M. Genkin, *Fiz. Tverd. Tela* **3**, 2097 (1961); see *Soviet Phys.—Solid State (English Transl.)* **3**, 1523 (1962); see erratum in *Sov. Phys. Solid State* **4**, 1019 (1962).
[261] W. Martienssen, *Phys. Chem. Solids* **2**, 257 (1957).
[262] A. Gold and R. S. Knox, *J. Chem. Phys.* **36**, 2805 (1962).

respectively, to values of 80×10^{-12} and (using $|\mathbf{K}| = 10^7$ cm^{-1}) 0.8×10^{-12} sec for the direct exciton in Ge, compared with the value of 4 to 8×10^{-12} sec derived from the line width.[263]

The contributions of optical phonons to line broadening are also given in the papers of Ansel'm and Firsov and Toyozawa, but we will not go into them here. Long-wavelength polar modes are less effective as scatterers than acoustcial modes, since the (assumed) purely electrostatic interaction tends to be canceled out by the exciton's over-all neutrality. This may be seen in the coupling constant (10.23); in an expansion of $-q_h$ and q_e in powers of $\mathbf{K}' - \mathbf{K}$ (which is equal to the scattering phonon wave vector), the leading terms are -1 and $+1$, respectively. Scattering becomes appreciable when the phonon wavelength becomes short enough to be comparable to the dimensions of the exciton itself. Short-wavelength optical modes are probably important in the Cu_2O $2s$ broadening mentioned earlier, and are surely important in the processes known specifically as "indirect transitions" (Section 10d).

(2) *Toyozawa's General Theory*.[264] In order to examine the severity of the approximations involved in the use of simple perturbation theory, Toyozawa[27] developed a formalism which, in principle, handles $\mathscr{H}_{eL}$ at any strength but which is still limited to first order in $\mathscr{H}_{eR}$. He computes the wave function of the system at time t by first using $\mathscr{H}_{eR}$ as a perturbation:

$$\Psi(t) = e^{-i\mathscr{H}t/\hbar} \Psi(0)$$

$$= e^{-i(\mathscr{H}_0 + \mathscr{H}_{eL} + \mathscr{H}_L + \mathscr{H}_R)t/\hbar} \left[1 + (i\hbar)^{-1} \int_0^t dt_1 \mathscr{H}_{eR}(t_1) + \cdots \right] \Psi(0). \quad (10.32)$$

Here $\mathscr{H}_{eR}(t_1)$ stands for $\mathscr{H}_{eR}$ in the interaction representation, and $\Psi(0)$ is a state in which there is a radiation field present, no excitons, and an equilibrium phonon distribution. Thus in our usual notation $\Psi(0) = |\cdots n_{\xi\kappa} \cdots; 0; \cdots n_{\sigma f} \cdots\rangle$. The inverse lifetime of a photon $\xi\kappa$ is defined as the thermal average of the probability, per $\xi\kappa$ photon and

---

[263] S. Zwerdling, L. M. Roth, and B. Lax, *Phys. Chem. Solids* **8**, 397 (1959).

[264] Other relevant articles are the following: A. S. Davydov and E. I. Rashba, *Ukr. Fiz. Zh.* **2**, 226 (1957) (weak coupling); E. I. Rashba, *Opt. i. Spektroskopiya* **2**, 568 (1957) (strong coupling); W. Biem, *Z. Physik* **164**, 199 (1961) (weak coupling); W. Weller, *Z. Naturforsch.* **16a**, 401 (1961) (strong coupling).

per unit time, that a state $\Psi_f(t) = | \cdots n_{\xi\kappa} - 1 \cdots; 1_{\nu\mathbf{K}}; \cdots n'_{\sigma\mathfrak{f}} \cdots \rangle$ has been realized. Thus

$$\frac{1}{\tau(\xi\kappa)} = \frac{1}{tn_{\xi\kappa}} |\langle \Psi_f(t) | \Psi(t)\rangle|^2$$

$$= \frac{1}{\hbar^2} \sum_{\nu} \sum_{\nu'} g_\nu(\xi\kappa)^* g_{\nu'}(\xi\kappa) \int_{-\infty}^{\infty} dt\, e^{i(\hbar\omega - E_{\nu\mathbf{K}})t/\hbar} \times \sum_n U_n(t; \nu\nu', \kappa) \quad (10.33)$$

where g is the exciton-photon coupling constant and

$$U_n(t; \nu\nu', \kappa) = \frac{1}{(i\hbar)^n} \int_0^t dt_1 \cdots \int_0^{t_{n-1}} dt_n$$

$$\times \frac{\text{Tr } e^{-\mathscr{H}_L/kT} (\nu\kappa | \mathscr{H}_{eL}(t_1) \cdots \mathscr{H}_{eL}(t_n) | \nu'\kappa)}{\text{Tr } e^{-\mathscr{H}_L/kT}}. \quad (10.34)$$

The several missing steps in this derivation may be found in Toyozawa's paper.[27] The absorption coefficient is defined as the penetration depth, or

$$\mu^{-1} = \text{(photon velocity)} \times \text{(photon lifetime)},$$

$$\mu^{-1} = (c/n)\, \tau(\xi\kappa) \quad (10.35)$$

and it reduces to the one derived in Section 9 when $\mathscr{H}_{eL}$ is vanishingly small. The use of c/n as the photon velocity in Eq. (10.39) is consistent with the use of $n^2 \mathsf{E}^2/4\pi$ as the energy density in the derivation of the coupling constant g (cf. Section 8a). When $\mathscr{H}_{eL}$ goes to zero, all the U_n are zero except U_0, which is simply $\delta_{\nu\nu'}$. We obtain

$$\mu(\omega) = \frac{n}{c\hbar^2} |g_\nu(\xi\kappa)|^2 \int_{-\infty}^{\infty} dt\, e^{i(\hbar\omega - E_{\nu\mathbf{K}})t/\hbar}. \quad (10.36)$$

Observing that the integral is just $2\pi\delta(\omega - E_{\nu\mathbf{K}}/\hbar)$ and substituting the value (9.41) of g, we regain the expression (9.11) for μ.

The general formalism handles as well as intraband scattering. Let us consider, however, some specific intraband results in the limits of weak and strong coupling for which, under certain simplifying assumptions, Toyozawa is able to sum the series in Eq. (10.33). The most severe assumption is that the density of exciton states is smooth at $\mathbf{K} = 0$; it is so severe that "weak coupling" itself must be defined in terms of the smoothness parameter as well as interaction energies. Nevertheless, when weak coupling does obtain, a Lorentzian absorption curve results:

$$\mu_{\text{weak}}(\omega) = \frac{n}{c\hbar^2} \, |g_\nu(\xi\varkappa)|^2 \, \frac{\hbar^2 \Gamma_{\nu\mathbf{K}}}{[\hbar\omega - (E_{\nu\mathbf{K}} + \Delta_{\nu\mathbf{K}})]^2 + \frac{1}{4}(\hbar\Gamma_{\nu\mathbf{K}})^2}, \quad (10.37)$$

where $\hbar\Gamma_{\nu\mathbf{K}}$ and $\Delta_{\nu\mathbf{K}}$ are the breadth and shift of the state $\Psi_{\nu\mathbf{K}}$ computed above by perturbation theory. When strong coupling exists, one obtains a Gaussian curve:

$$\mu_{\text{strong}}(\omega) = \frac{n}{c\hbar^2} \, |g_\nu(\xi\varkappa)|^2 \, \frac{1}{\pi^{1/2} D_\nu} \exp\left(-\frac{(\hbar\omega - E_{\nu\mathbf{K}})^2}{2D_\nu^2}\right) \quad (10.38)$$

of half-width $2^{3/2} \log 2 D_\nu$, where

$$D_\nu^2 = \sum_\sigma \sum_\mathbf{f} |G_{\sigma\mathbf{f}}(\nu\mathbf{K} + \mathbf{f}, \nu\mathbf{K})|^2 (2\bar{n}_{\sigma\mathbf{f}} + 1).$$

Because of the form of most coupling constants (Section 10a), D_ν is independent of $\mathbf{K}$. When only one phonon (of frequency ω') is important, D_ν^2 has the familiar form $D_\nu^2 = G'^2 \coth(\hbar\omega'/2kT)$. Toyozawa's strong coupling limit is attained when D_ν is much larger than the exciton band width, and describes a situation in which the exciton essentially becomes trapped at once.

Although these tractable limiting cases make it appear that the only job left is to fill in the intermediate coupling case, it turns out that even the weak coupling result will not always be uself in practice. Referring to Fig. 27a, consider temperatures at which scattering takes place to states well up into the exciton band. Clearly no absorption is possible at energies equally spaced *below* the level at $\mathbf{K} = 0$, because there are no states there. This makes the Lorentzian curve untenable. It has, of course, long been known that an asymmetry would appear in an absorption line arising from an exciton band whose minimum is at $\mathbf{K} = 0$, and the point at which Toyozawa's theory breaks down is in the assumption that the density of states is smooth in that region. He has recently extended the theory to account for this difficulty,[245] finding that multiphonon processes are very important near the region of a singular density-of-states derivative. His new line widths are proportional to T^2, and in the case of a band extremum at $\mathbf{K} = 0$, the long-wavelength absorption edge drops sharply to zero at an energy of about $\frac{1}{4}$ the half-width below the peak.

(3) *Other Topics.* The sharp cutoff which one imagines quali-

tatively (and which Toyozawa actually computes) is alleviated by "retardation effects," which in the present context means treating $\mathcal{H}_{eR}$ as a nonsmall perturbation. As we have seen in Section 9c, there *are* states below the usual exciton band, and their coupling to phonons will be quantitatively, if not qualitatively, different from that to a simple **K** = 0 exciton. A calculation along these lines has been done by Agranovich and Konobeev.[265] Their calculation, applied to the case of Frenkel-acoustic coupling, yields results which appear identical to those given here, with two exceptions. The exciton energy is replaced by the energy of the polariton, and the absorption coefficient must be defined in terms of an energy velocity consistent with the energy density of the carrier waves. They therefore use the polariton group velocity in place of c/n. Their results are valid only in the long wavelength tail and will be discussed in Section 10c.

One of the most popular asymmetries in line shape is that of the $n = 2$ line in Cu_2O (see Fig. 2). We mention here the various possible causes which have been suggested for this asymmetry. Zverev and co-workers[266] find a good fit to a modified Lorentzian curve originally proposed by Toyozawa[27]; the modification arises, in this case, from a weak variation in the density of exciton states at the peak energy. Elliott,[258] on the other hand, considers it likely that the asymmetry is due to a variation with energy in the exciton-phonon coupling constant. Hopfield,[267] finally, considers the case of absorption to a single final state by way of two different types of intermediate state, both of which may contribute appreciably. He points out that absorption coefficients are not additive in this case, and that with appropriate phase relations between matrix elements, the background upon which the $n = 2$ line is superposed may affect the wings of that line in such a way as to cause the observed asymmetry. All three of these possible explanations seem plausible, and all three effects are probably present. One would be harder pressed, it would seem, to explain the line shape if it were perfectly symmetric.

c. Line Shifts and Urbach's Rule

While the most interesting thermal effect on direct transitions is

[265] V. M. Agranovich and Yu. V. Konobeev, *Fiz. Tverd. Tela* **3**, 360 (1961); see *Soviet Phys.—Solid State (English Transl.)* **3**, 260 (1961).

[266] L. P. Zverev, M. M. Noskov, and M. Ya. Shur, *Fiz. Tverd. Tela* **2**, 2643 (1960); see *Soviet Phys.—Solid State (English Transl.)* **2**, 2357 (1961).

[267] J. J. Hopfield, *Phys. Chem. Solids* **22**, 63 (1961).

seen in line shapes, there is usually an over-all shift in position which can be attributed to phonons. It is generally observed that exciton absorption lines shift toward lower energies as the temperature is raised. An interesting counterexample is TlCl, part of whose absorption spectrum[230] is shown in Fig. 28. At normal temperatures

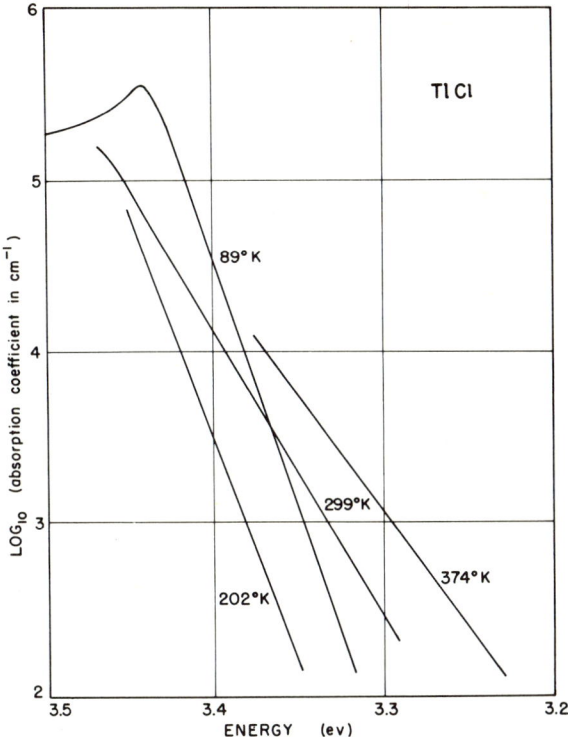

FIG. 28. Anomalous temperature dependence of the TlCl absorption edge [after S. Tutihasi, *Phys. Chem. Solids* **12**, 344 (1960)]. The absorption is plotted logarithmically to illustrate an Urbach-rule edge [Eq. (10.40)].

($\gtrsim 150°$ K), the edge (which probably follows the peak) moves toward lower energy. This is "normal" behavior. However, at low temperatures the motion reverses.

Theories constructed in connection with ordinary band theory indicate that the gap energy E_G decreases as T rises. Although the effect of an increasing lattice constant would presumably be to increase E_G, because the valence and conduction bands would grow narrower, this is not always the case; anyway, two much larger effects which decrease E_G dominate. They are the broadening and shifting of the one-electron states, specifically those states at the band edges, due to interactions with phonons.[268] On the Wannier model, since exciton energy levels are so intimately related to E_G, one qualitatively expects them to share its temperature dependence. This is reflected in the perturbation expression (10.29) for $\Delta_{\nu\kappa}$, which for the Wannier exciton is related to Fan's result[268] and is surely negative for low-lying exciton states; in one example, the following empirical form of E_G in Cu_2O between 4.2 and 77° K has been deduced[269] from the behavior of the yellow series:

$$E_G(T) = 2.171\,(1 - 1.87 \times 10^{-5}\,T - 3.5 \times 10^{-7}\,T^2)\,\text{ev}. \qquad (10.39)$$

Aside from the perturbation results presented in the previous section, few calculations designed to deal specifically with temperature shifts of exciton peaks have appeared. Von Hippel[8,270] applied thermodynamic arguments to the simplified transfer model in polar crystals with somewhat inconclusive results, and Ohta[271] recast the Fan calculation in terms of Frenkel exciton shifts. Another class of calculations which can lead to a temperature dependence of line positions are those of Section 7 dealing with the general clothed-particle theory of exciton structure. In some cases a temperature-dependent $1s$ energy results,[272] but usually no acoustic phonon interactions have been accounted for in these theories. We can conclude that line shifts are not at all understood quantitatively, and no explanation for anomalous blue shifts is available.

Much more interesting is that the line shift problem is an empirical relationship between the absorption coefficient and the absolute temperature first discovered in silver halides by Urbach[273]:

$$\mu = \mu_0 \exp\,[\sigma(\hbar\omega_0 - \hbar\omega)/kT]. \qquad (10.40)$$

---

[268] A. Radkowsky, *Phys. Rev.* **73**, 749 (1948); H. Y. Fan, *ibid.* **82**, 900 (1951).

[269] S. Nikitine, G. Perny, and M. Sieskind, *Compt. Rend.* **238**, 1987 (1954).

[270] A brief review of this work may be found in H. Haken, *Fortschr. Physik* **6**, 271 (1958).

[271] T. Ohta, *Progr. Theoret. Phys. (Kyoto)* **10**, 472, 474 (1953).

[272] See, e.g., I. G. Zaslavskaya, *Fiz. Tverd. Tela* **3**, 2240 (1961); see *Soviet Phys.— Solid State (English Transl.)* **3**, 1627 (1962).

[273] F. Urbach, *Phys. Rev.* **92**, 1324 (1953).

Here $\hbar\omega_0$ is a constant of the order of magnitude of the peak absorption energy, and σ is a constant of order unity. This rule holds only for $\omega < \omega_0$, but is remarkably accurate over several decades in $\mu(\omega)$ and over a fairly wide temperature range, as shown in Table IV and in

TABLE IV. SOLIDS EXHIBITING ABSORPTION SPECTRA OBEYING URBACH'S EMPIRICAL RULE [EQ. (10.40)]

The symbols are defined and discussed in connection with this equation.

Substance	$\hbar\omega_0$ (ev)	σ	Valid range	Reference
AgBr	2.81	1.0	10^{-2} to 30 cm^{-1}; 100-650° K	273,274
AgCl	3.33	0.82	10^{-2} to 30 cm^{-1}; 100-650° K	274
CdS	Various	2.2	20 to 10^4 cm^{-1}; 90-340° K	86
KBr	6.80	0.79	10^{-2} to 10^4 cm^{-1}; 80-1000° K	275,276
KCl	7.76	0.80	1 to 200 cm^{-1}; 273-500° K	275
KI	5.9	0.79	5 to 7000 cm^{-1}; 300-900° K	276,277
PbI$_2$	2.53	1.3	10 to 1000 cm^{-1}; 77-300° K	91
Se (amorph.)	1.8	0.4	3-400 cm^{-1}; 140-300° K	278
TlCl	3.5	0.5	0.1 to 10^4 cm^{-1}; 20-150° K	230
		1.1	0.1 to 10^4 cm^{-1}; over 150° K	

Fig. 28 for TlCl. For these reasons it cannot be waved aside as a fortuitous approximation to a Gaussian or Lorentzian curve, and at present no completely satisfactory explanation for it exists. We mention four relevant theories. Dexter[232] approached the problem by computing the effect, on the absorption spectrum for band-to-band transitions, of a lattice deformation Δ due to lattice vibrations which manifests itself as a local disturbance of the band gap,

$$E_G(\Delta) = E_G(0) - E_1\Delta.$$

[274] F. Moser and F. Urbach, *Phys. Rev.* **102**, 1519 (1956); F. Moser, private communication.
[275] K. Kobayashi and T. Tomiki, *Phys. Chem. Solids* **22**, 73 (1961).
[276] W. Martienssen, *Phys. Chem. Solids* **8**, 294 (1959).
[277] U. Haupt, *Z. Physik* **157**, 232 (1959).
[278] M. A. Gilleo, *J. Chem. Phys.* **19**, 1291 (1951). Prof. R. M. Blakney kindly called the author's attention to this work and estimated the constants given in Table IV.

The band gap is thus statistically distributed according to the probability that a deformation Δ is realized, which is given by the statistical factor $P(\Delta) = \text{const} \exp(-B\Delta^2/kT)$, where B is some average elastic constant. Optical absorption is assumed to set in at an energy $E_G(\Delta)$ with a characteristic square-root dependence. Thus μ is taken to be the following average (the expression in the square bracket is defined as zero for negative argument):

$$\mu(\omega) = \text{const.} \frac{\int_{-\infty}^{\infty} e^{-B\Delta^2/kT}[\hbar\omega - E_G(0) - E_1\Delta]^{1/2}\, d\Delta}{\int_{-\infty}^{\infty} e^{-B\Delta^2/kT}\, d\Delta}. \quad (10.41)$$

This quantity, which Dexter has computed,[232] behaves like Urbach's rule over ranges of temperature and absorption coefficient much smaller than those observed. Toyozawa[244,279] has noted that the quantity

$$\mu(\omega) = \text{const.} \frac{\int_{-\infty}^{\infty} e^{-B\Delta^2/kT}\, \delta(\hbar\omega - E_0 - E_1\Delta - E_2\Delta^2)\, d\Delta}{\int_{-\infty}^{\infty} e^{-B\Delta^2/kT}\, d\Delta} \quad (10.42)$$

where E_0 is the energy of a single sharp absorption line in the absence of any dilatation, reproduces Urbach's rule over the whole range if E_1 is neglected. This is clearly unacceptable, as he points out, since E_2 must be taken to be several ev in order to obtain a reasonable value of the constant σ in practical cases, while it is known that E_1 also must be of the order of several ev to explain line broadening in the main peak in the same solids. Toyozawa believes that this indicates that different phonons are responsible for the tail absorption from those which provide the main broadening, and suggests that optical phonons may cause the tail.

The foregoing mechanism for Urbach's rule is probably physically sound, and its failure to lead to quantitatively valid results can be explained as follows. First, the fine details of phonon interactions are completely smoothed over by the use of a simple dilatation for the lattice distortion, and second, neither of the functional forms averaged in Eqs. (10.41) and (10.42) actually corresponds, precisely,

[279] Y. Toyozawa, *Progr. Theoret. Phys. (Kyoto)* **22**, 445 (1959).

to a predicted absorption spectrum for $\varDelta = 0$ [see, e.g., Eq. (9.28) for the case of the continuum spectrum]. Replacing the delta function in Eq. (10.42) with a Gaussian or Lorentzian shape does not appreciably alter Toyozawa's result, and replacing the square root in (10.41) by Elliott's more complete band shape would be meaningless unless some discrete absorption lines were also included in the manner of (10.42).

One of the basic problems obstructing progress on Urbach's rule is the need for realizing states far below the direct-transition exciton energy. As we have seen earlier, the polariton (mixed exciton-photon) does provide such states, and this brings us to the third relevant calculation. In the work of Agranovich and Konobeev mentioned just above,[265] absorption is found to tail off to energies considerably below the usual exciton band edge, but its frequency dependence differs considerably from Urbach's rule for a model crystal using Frenkel-acoustic phonon coupling. We believe this approach is worth pursuing in at least two ways: inclusion of second-order processes which scatter the polaritons back to nearly their original state by two-optical-phonon transitions; and recomputing with first-order theory but with different coupling constants. The qualitative arguments given recently by Hopfield[267] are based on the fact that an incident photon creates a polariton state which may thermally ionize at a rate proportional to a Boltzmann factor similar to Urbach's, with $\sigma \equiv 1$. Finally, a totally different approach has been taken by Redfield,[280] who has studied the effect on the absorption edge of the large random electric fields produced by impurities in a semiconductor. The effect is computed on the basis of the mechanism of Franz and Keldysh,[281] and leads to exponential tails in the absorption edge which, Redfield believes, will also be the case when the fields involved are those produced by phonons.

Another means of realizing states below the direct edge is by the presence of lower *indirect* exciton states, to be discussed directly below. Again, the predicted frequency dependence of absorption is nothing like Urbach's rule, and enough concrete examples of indirect transitions are known that that theory can hardly be considered suspect. The rule remains a challenge to theorists as of the present

---

[280] D. Redfield, *Phys. Rev.* **130**, 916 (1963).

[281] W. Franz, *Z. Naturforsch.* **13a**, 484 (1958); L. V. Keldysch, *Zh. Eksperim. i Teor. Fiz.* **34**, 1138 (1958) [*English Transl.: Soviet Phys. J.E.T.P.* **7**, 788 (1958)].

158   III. ABSORPTION AND DISPERSION OF LIGHT

writing. Further complicating the picture is the appearance of Urbach-rule absorption in impurity centers[275,282] and amorphous systems.[278]

d. Indirect Transitions[19,21]

It is often implicitly assumed, in the case of solids where no information to the contrary is available, that the lowest exciton state occurs at $\mathbf{K} = 0$. This assumption was in fact almost universal until the cyclotron-resonance discovery that two very important solids (Ge and Si) possess conduction band minima at points away from the center of the BZ which, moreover, do not correspond to the positions of the top of the valence band. The lowest electronic states were then seen to be hole-electron pairs of momenta $(0, \mathbf{k}_c)$, where $|\mathbf{k}_c|$ is quite large; in order that a photon excite this state, an additional momentum $\mathbf{k}_c$ is required, and it is provided[213] by annihilation of a phonon of momentum $\mathbf{k}_c$ or creation of a phonon of momentum $-\mathbf{k}_c$. Precisely the same momentum considerations apply to the exciton, since (Section 4) an exciton constructed from bands whose extrema differ by an amount $\mathbf{K}_0$ will have a momentum $\mathbf{K}_0$ which must be supplied by the phonon field during an optical transition. Just as the detailed absorption edge shape (9.28) for direct transitions differs considerably from the one predicted by band theory, so does the shape for indirect transitions. Prompted by detailed data provided by the experimental group at Malvern,[97,98] Elliott[21] developed a theory of indirect exciton transitions which has provided a basis for the interpetation of edge spectra in a growing number of crystals. There has recently appeared a very comprehensive review of the theory[19] and its application to Ge and Si, so we shall be content with a qualitative exposition and a survey of applications to other crystals.

With caution, we superpose exciton bands (dashed lines) on a one-electron band picture of Ge (Fig. 29). Band i represents exciton states constructed from holes near $\mathbf{k} = 0$ and electrons near $\mathbf{k} = \mathbf{K}_0$. Band d represents "direct" exciton states constructed from holes and electrons both near $\mathbf{k} = 0$. The circle represents the ground state Ψ_0 of the crystal. We now consider the possible kinds of transitions from Ψ_0 to the vicinity of the minimum of band i, two kinds of which are sketched as arrows in the figure. The upper arrow terminates at a state of energy $E_{i\mathbf{K}_0} + \hbar\omega_{\sigma,\mathbf{K}-\mathbf{K}_0}$. A photon of this energy and essen-

[282] H. Mahr, *Phys. Rev.* **122**, 1464 (1961); **125**, 1510 (1962).

tially zero momentum $\varkappa$ has created the exciton state $\Psi_{i\mathbf{K}_0}$ and a phonon in mode σ, $\varkappa - \mathbf{K}_0$. Similarly, the lower arrow represents the creation of $\Psi_{i\mathbf{K}_0}$ and the annihilation of a σ, $\mathbf{K}_0 - \varkappa$ phonon by a photon of energy $E_{i\mathbf{K}_0} - \hbar\omega_{\sigma,\mathbf{K}_0-\varkappa}$.

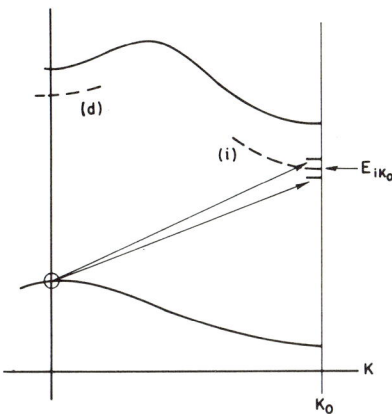

FIG. 29. One-electron bands and associated exciton bands in a typical crystal exhibiting indirect transitions. The arrows indicate two possible phonon-assisted transitions. The circle and dashed lines indicate the ground state and low-lying exciton states of the crystal, positioned arbitrarily (but conveniently) with reference to the top of the valence band. $\mathbf{K}_0$ marks the edge of the BZ, and (d) and (i) designate the "direct" and "indirect" portions of the lowest exciton band.

Transition probabilities for these processes have actually been given in Section 10b; we constructed first-order states $\Psi_{i\mathbf{K}_0}^{(1)}$ using the exciton-phonon interaction, and then computed the radiative transition probability from Ψ_0 to the first-order states. The result, given as Eq. (10.27), was discussed as a line-broadening phenomenon. Here, the "central line" is missing, and Eq. (10.27) gives the entire absorption spectrum. Furthermore, here we are not troubled in general by vanishing energy denominators because the intermediate states $\Psi_{\nu'\varkappa}$ are not necessarily anywhere near the absorption energy under consideration. One such intermediate state, for example, might be the direct state $\Psi_{d\varkappa}$; another might be a direct exciton constructed from holes and electrons both near $\mathbf{K}_0$.

Consider, now, what is to be expected qualitatively of the absorption spectrum in the region of $\hbar\omega \sim E_{i\mathbf{K}_0}$. We have mentioned above only one possible pair of transitions. Had we considered *any* point in the

i exciton band in the vicinity of $\mathbf{K}_0$, we could have made precisely the same argument, because it is always possible to find an appropriate phonon to conserve momentum. Thus, as soon as the incident photon energy has reached $E_{i\mathbf{K}_0} - \hbar\omega_{\sigma\mathbf{K}_0}$, a continuum absorption due to simultaneous phonon absorption will set in, as shown in Fig. 30.

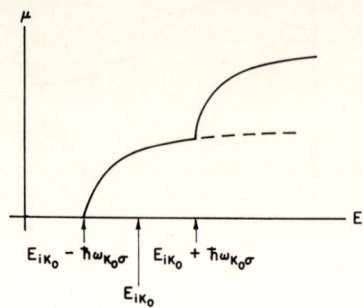

Fig. 30. Double step characteristic of an indirect absorption edge. The step at lower energy corresponds to phonon absorption simultaneously with exciton production. This step disappears at low temperature, i.e., when no phonons are present to be absorbed.

When $E_{i\mathbf{K}_0} + \hbar\omega_{\sigma\mathbf{K}_0}$ is reached, another continuum, due to simultaneous creation of $\sigma\mathbf{K}_0$ phonons along with the excitons, sets in. The shape of these continua may be deduced as follows. The phonons participating in the interaction come from a fairly limited region of the BZ centered at $\mathbf{K}_0$, so all those from the same branch have roughly the same energy. For the same reason, it makes sense to assume that their coupling between the intermediate state $\Psi_{\nu'\mathbf{\kappa}}$ and the final state is independent of $\mathbf{K}$. Hence, if the intermediate state is not too close to the final state, the quantities in absolute value signs in Eq. (10.27) are independent of $\mathbf{K}$ and equal to a constant ($\propto D\,|\,\phi'(0)\,|^2$ in Elliott's original notation) which depends on the type of phonon involved. Thus the dependence of Eq. (10.27) on energy, taking $\varkappa = 0$, is

$$\sum_\sigma \text{const}_\sigma \sum_\mathbf{K} \{[\bar{n}_{\sigma\mathbf{K}}(T) + 1]\,\delta[\hbar\omega - (E_{i\mathbf{K}} + \hbar\omega_{\sigma\mathbf{K}})]$$
$$+ \bar{n}_{\sigma\mathbf{K}}(T)\,\delta[\hbar\omega - (E_{i\mathbf{K}} - \hbar\omega_{\sigma\mathbf{K}})]\}. \qquad (10.43)$$

Similarly, $\bar{n}_{\sigma \mathbf{K}}(T)$ is essentially **K**-independent; and assuming a simple exciton band centered at $\mathbf{K}_0$, with

$$E_{i\mathbf{K}} = E_G - G_i + \frac{\hbar^2(\mathbf{K} - \mathbf{K}_0)^2}{2M_i^*},$$

we have

$$\mu(\omega) = \sum_\sigma C_\sigma \{[\bar{n}_{\sigma \mathbf{K}_0}(T) + 1] (\hbar\omega - E_G - G_i - \hbar\omega_{\sigma \mathbf{K}_0})^{1/2}$$
$$+ \bar{n}_{\sigma \mathbf{K}_0}(T) (\hbar\omega - E_G - G_i + \hbar\omega_{\sigma \mathbf{K}_0})^{1/2}\}. \quad (10.44)$$

E_G is of course the indirect gap and G_i is the binding energy of the indirect exciton. Equation (10.44) is the function sketched, for a single phonon type, in Fig. 30. A particularly significant feature of such a spectrum is the disappearance, at low temperatures, of the low-energy step. Physically, the phonons necessary for the step are simply not present. The low-temperature edge of AgCl is shown in Fig. 31.

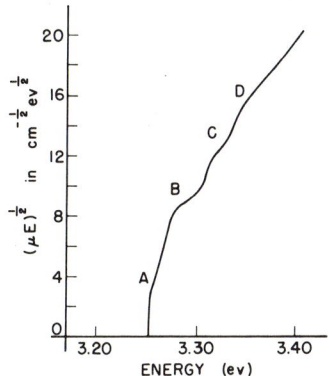

FIG. 31. Steps in the absorption edge of AgCl [after F. C. Brown, T. Masumi, and H.H. Tippins, *Phys. Chem. Solids* **22**, 101 (1961)] at 4.2° K. These steps are phonon *emission* steps at this temperature, and indicate the possible presence of several exciton bands. The quantity $(\mu E)^{1/2}$ was plotted as ordinate to facilitate comparison with the theory of indirect band-to-band transitions.

In the case of indirect band-to-band transitions, it was known[213] that the absorption had the form (10.44) with the square roots replaced by squares. (This change in energy dependence can be seen to be due to the change in the density of final states available to the system.) Thus, excitons change the anticipated edge shape radically.

In Elliott's theory, the square dependence is also found when $\Delta E = \hbar\omega - E_G \gg G_i$, the usual condition for validity of the free-pair approximation. McLean[19] discusses this limiting case as well as others not treated here, such as the one in which the important transitions are "forbidden."

Setting aside for the moment the fact that the above electronic model is oversimplified for most practical cases, we note that according to Eq. (10.44), one interprets an indirect spectrum by looking for a set of superposed double steps. The separation between steps is twice the energy of the σ phonon at $\mathbf{K}_0$, so there are rich possibilities for interrelating band structure and phonon spectra obtained from widely different studies. From a known band structure one can deduce from symmetry arguments the types of phonons which should contribute to Eq. (10.44) and the corresponding values of $2\hbar\omega_{\sigma\mathbf{K}_0}$ are obtainable from neutron spectroscopy or normal mode calculations.

In Ge, four separate phonon energies have been deduced from the spectrum, and they are in extremely good agreement with neutron scattering data. Furthermore, two excitons whose separation is 0.0010 ev are seen in interaction with longitudinal acoustical (LA) phonons; they are to be interpeted[283] not as $n = 1$ and $n = 2$ Wannier states, but as two $n = 1$ states arising as different combinations of the several possible $n = 1$ states constructed at equivalent conduction band minima and with different branches of the valence band. The situation in Si is not as clear-cut, but two phonons (TA and TO) have been found and correlated with neutron spectra.

A final important example of indirect transitions is again found in Cu_2O. We remarked above that indirect transitions to large values of $\mathbf{K}_0$ can be understood on the basis of the general ideas of line broadening discussed in Section 10b, with considerable simplification because of the absence of a "central line" and of small energy denominators. The $1s$ quadrupole line in Cu_2O (at $\mathbf{K}_0 = 0$) is essentially absent because of its tiny oscillator strength, and a double edge is found on each side of it as the temperature is raised.[284,285] It therefore would appear[258] that some kind of optical phonons are the only ones contributing to indirect processes at this edge, because the assumptions

---

[283] T. P. McLean and R. Loudon, *Phys. Chem. Solids* **13**, 1 (1960).

[284] I. Pastrnyak, *Fiz. Tverd. Tela* **1**, 970 (1959); see *Soviet Phys. Solid State (English Transl.)* **1**, 888 (1959); I. S. Gorban and V. B. Timofeev, *Fiz. Tverd. Tela* **3**, 3584 (1961); see *Soviet Phys.—Solid State (English Transl.)* **3**, 2606 (1962).

[285] J. B. Grun, M. Sieskind, and S. Nikitine, *Phys. Chem. Solids* **19**, 189 (1961).

made about constant energies and coupling constants in deriving Eq. (10.44) are not applicable to acoustic modes at $\mathbf{K}_0 = 0$.

Indirect transitions have been studied and interpreted along the lines of this section in diamond,[286] SiC,[99] AgCl and AgBr,[100] and CdS (near $\mathbf{K}_0 = 0$),[88] as well as in Ge and Si.[19]

11. Anomalous Waves and Spatial Dispersion

In this brief section we present a simplified version of a subject which has excited considerable interest in the Soviet journals but which has not received any attention at all in the rest of the solid state literature. It concerns the possibility that two different kinds of waves of the same energy and same polarization can exist in a crystal. These waves would differ only in wave vector, i.e., in their index of refraction. The one with an anomalously high wave vector is an "anomalous wave." We shall now examine the extent to which these waves can be regarded as photons.

The real part of the refractive index in isotropic crystals satisfies the following general relationship which is obtainable from Eq. (8.5):

$$n^4 - n^2\epsilon - 4\pi\sigma^2\omega^{-2} = 0. \tag{11.1}$$

Of the four solutions of this equation as it is commonly used, two are imaginary and do not lead to propagating waves. The other two are negatives of each other and describe physically identical waves propagating in opposite directions in the medium. Thus the refractive index would appear to be a unique function of the frequency, determined by the dependence of $\epsilon(\omega)$ and $\sigma(\omega)$ on ω. The Lorentz model provides specific forms (8.16) and (8.17) for these functions.

In 1957, Pekar[287] observed that in the treatment of dispersion by exciton theory, this relatively simple situation would be modified if the frequencies of the normal modes of the system themselves were allowed to depend on the wave vector, as they do in exciton theory. For then ω_0 in Eqs. (8.16) and (8.17) would be replaced by $\omega_0(k) = \omega_0(n\omega/c)$, and the following equation, of much higher order in n, would result:

$$n^4 - n^2\epsilon(\omega, n) - 4\pi\sigma(\omega, n)^2 \omega^{-2} = 0. \tag{11.2}$$

[286] C. D. Clark, *Phys. Chem. Solids* **8**, 481 (1959).

[287] S. I. Pekar, *Zh. Eksperim. i Teor. Fiz.* **33**, 1022 (1957); see *Soviet Phys. J.E.T.P. (English Transl.)* **6**, 785 (1958); *Zh. Eksperim. Teor. Fiz.* **34**, 1176 (1958), or *Soviet Phys. J.E.T.P. (English Transl.)* **7**, 813 (1958); *Phys. Chem. Solids* **5**, 11 (1958).

[In simple cases, one might have $\omega_0(k) = \omega_0(0) + \text{const } (n^2\omega^2/c^2)$.] Pekar accordingly found new solutions for the index of refraction, calling the electromagnetic waves corresponding to these new solutions "anomalous waves." As it happened, Pekar made these observations in the context of exciton theory at about the time that interest was being generated in the exciton-photon, strong-coupling problem ("retardation effect"), and the two effects have on occasion been confused in the literature. Pekar's new solutions are not a *result* of the strong coupling; they occur whenever there is any *curvature* of the ordinary exciton band in the region of large exciton-photon coupling, as indicated, in a somewhat exaggerated form, in Fig. 32. Pekar's

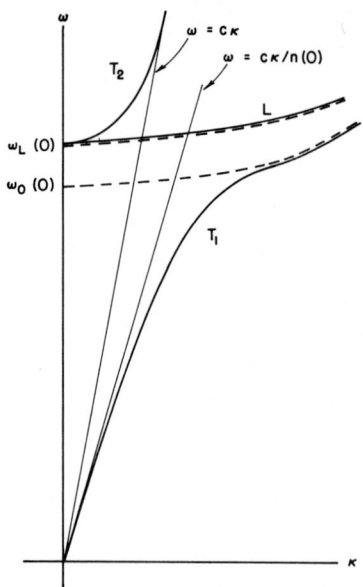

FIG. 32. Illustrating the possible changes in Fig. 21 which follow upon the introduction of coupling between Lorentz oscillators. A true *band* of energies $\omega(\kappa)$ exists, and the resulting curvature makes possible the "anomalous waves" (see text).

observation applies to the Lorentz model, as well as to quantum-mechanical models, whenever the purely mechanical normal modes of the system display dispersion, i.e., a dependence of frequency on wave number. This will be the case when any coupling (such as a dipolar coupling[288]) between the individual oscillators is introduced,

[288] See e.g., U. Fano, *Phys. Rev.* **103**, 1202 (1956); **118**, 451 (1960).

11. ANOMALOUS WAVES AND SPATIAL DISPERSION

just as two-center matrix elements between localized excitations lead to a nonzero exciton band width. The Pekar waves can therefore be discussed in terms of the Lorentz model, avoiding the notational problems peculiar to the large Russian literature which now exists on the subject,[48,289-297] and, incidentally, avoiding the problem of relating the effect to "retardation," since the latter is already a part of the Lorentz theory.

To simplify our discussion we consider the case of vanishing damping, and assume that the frequencies of interest are in the vicinity of one of the oscillator frequencies ω_{0j}. Then the contribution $\epsilon'(\omega)$ of the remaining oscillators to the dielectric function is nearly constant, and we can write

$$\epsilon = \epsilon' + \frac{f_j \omega_P^2}{\omega_{0j}^2 - \omega^2} \approx \epsilon' + \frac{f_j \omega_P^2}{2\omega(\omega_{0j} - \omega)}. \tag{11.3}$$

Since $\sigma \equiv 0$, Eq. (11.2) reduces to $n^2 = \epsilon(\omega, n)$. However, if it is now assumed that $\hbar\omega_{0j}$ has the form

$$\hbar\omega_{0j} = \hbar\omega_{0j}(0) + (\hbar^2/2M^*) k^2 \tag{11.4}$$

where $k = n\omega/c$, Eq. (11.2) again becomes quadratic in n^2 because of the appearance of n in $\epsilon(\omega, n)$.

[289] V. L. Ginzburg, *Zh. Eksperim. i Teor. Fiz.* **34**, 1593 (1958); see *Soviet Phys. J.E.T.P. (English Transl.)* **7**, 1096 (1958).
[290] A. S. Davydov and A. F. Lubchenko, *Zh. Eksperim. i Teor. Fiz.* **35**, 1492 (1958); see *Soviet Phys. J.E.T.P. (English Transl.)* **8**, 1048 (1959).
[291] V. M. Agranovich and A. A. Rukhadze, *Zh. Eksperim. i Teor. Fiz.* **35**, 982 (1958); see *Soviet Phys. J.E.T.P. (English Transl.)* **8**, 685 (1959).
[292] A. F. Lubchenko, *Fiz. Tverd. Tela* **1**, 709 (1959); see *Soviet Phys.—Solid State (English Transl.)* **1**, 646 (1959).
[293] S. I. Pekar, *Zh. Eksperim. i Teor. Fiz.* **36**, 451 (1959); see *Soviet Phys. J.E.T.P. (English Transl.)* **9**, 314 (1959).
[294] S. I. Pekar and B. E. Tsekvava, *Fiz. Tverd. Tela* **2**, 261 (1960); see *Soviet Phys.—Solid State (English Transl.)* **2**, 242 (1960).
[295] S. I. Pekar, *Zh. Eksperim. i Teor. Fiz.* **38**, 1786 (1960); see *Soviet Phys. J.E.T.P. (English Transl.)* **11**, 1286 (1960).
[296] S. I. Pekar and V. L. Strizhevskii, *Fiz. Tverd. Tela* **2**, 894 (1960); see *Soviet Phys.—Solid State (English Transl.)* **2**, 816 (1960).
[297] B. E. Tsekvava, *Fiz. Tverd. Tela* **3**, 1164 (1961); see *Soviet Phys.—Solid State (English Transl.)* **3**, 847 (1961); *Fiz. Tverd. Tela* **4**, 501 (1962); see *Soviet Phys.—Solid State (English Transl.)* **4**, 364 (1962).

Its solutions are

$$n_\pm^2 = \tfrac{1}{2}\left[\frac{2M^*c^2}{\hbar\omega}\left(1 - \frac{\omega_{0j}}{\omega}\right) + \epsilon'\right]$$

$$\pm \left\{\tfrac{1}{4}\left[\frac{2M^*c^2}{\hbar\omega}\left(1 - \frac{\omega_{0j}}{\omega}\right) - \epsilon'\right]^2 + \frac{f_j M^* c^2 \omega_P^2}{\hbar\omega^3}\right\}^{1/2}. \quad (11.5)$$

For a positive mass M^*, the results appear as in Fig. 33. The dashed line shows the ordinary Lorentz-model expression for n^2, which according to Eq. (11.3) approaches $\epsilon' + f_j\omega_P^2\omega_{0j}^{-2}$ at low frequencies and ϵ' at high frequencies, and has a vertical asymptote at ω_{0j}. The new solutions approach the low- and high-frequency asymptotes but bot the vertical asymptote; rather, they approach the line

$$n^2 = 2M^*c^2\hbar^{-1}\omega^{-2}(\omega - \omega_{0j}).$$

Since Eq. (11.5) is derived using the approximate form of (11.3), and suffers particularly from the assumption of no damping, this is an oversimplified picture but it does show the appearance of two real indices of refraction at a given frequency. Quantitative data on Fig. 33

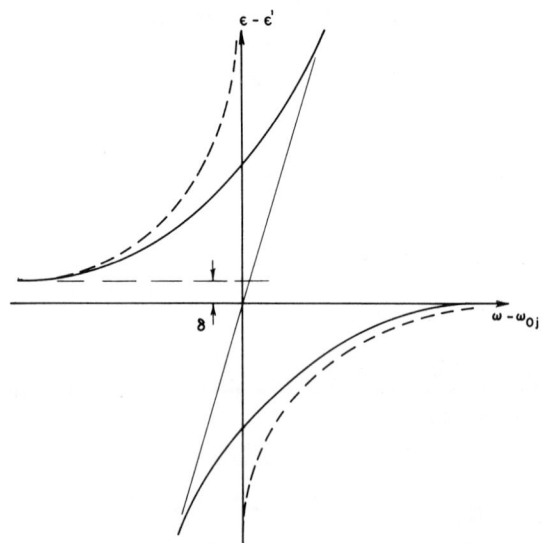

FIG. 33. General form of the solutions of Eq. (11.4) near $\omega = \omega_{0j}$ (solid lines). The slanted asymptote has a slope proportional to the assumed exciton mass M^*. The dashed lines are the solutions in the "normal" Lorentz case (infinite exciton mass). The distance δ is equal to $f_j\omega_P^2/\omega_0^2$.

and a similar plot for negative exciton mass are given in Pekar.[287,293] One may also infer the existence of Pekar's excitations by noting that, at certain energies, there exist two possible polaritons (Fig. 32).

Experiments have been performed[298,299] to search for Pekar's extra waves. They are based on the fact that waves of the same frequency but different indices n_+ and n_-, different polarizations $\mathbf{e}_+$ and $\mathbf{e}_-$, and different absorption coefficients μ_+ and μ_- traveling coherently through a crystal of thickness d exhibit an interference which results in a final intensity

$$I = I_+(\omega)\, e^{-\mu_+ d} + I_-(\omega)\, e^{-\mu_- d}$$
$$+ 2\mathbf{e}_+ \cdot \mathbf{e}_- [I_+(\omega)\, I_-(\omega)]^{1/2}\, e^{-1/2(\mu_+ + \mu_-)d} \cos\left[\frac{\Delta n(\omega)\, d}{c} + \alpha_{+-}\right] \quad (11.6)$$

where $\Delta n(\omega) = n_+(\omega) - n_-(\omega)$, I_+ and I_- are the original intensities, and α_{+-} the original phase difference of the waves. In the case of ordinary double refraction, $\mathbf{e}_+ \cdot \mathbf{e}_- = 0$ and the third term of Eq. (11.6) is ineffective. Here, we have a "self-interference" since the waves under consideration have the same polarization and the third term of Eq. (11.6) is nonzero. One thus expects an apparent optical density which depends on the thickness d in an oscillatory manner at a given frequency ω. In the experiments on anthracene[298] and Cu_2O,[299] such an oscillatory effect has been seen, and it is of quite a different order of magnitude in the two cases. It corresponds to $\Delta n = 50$ in anthracene and $\Delta n = 0.003$ in Cu_2O, if one merely assigns to the observed oscillations in O.D. a cosine dependence like the one in Eq. (11.6). This difference is consistent with the appearance of f_j in Eq. (11.5). In anthracene, the O.D. has been measured at fixed frequencies near an allowed transition as a function of d, while in Cu_2O an entire quadrupole absorption line shape has been measured as a function of d. In the latter case the peak absorption coefficient surely has an oscillatory dependence but it is not clear from the data that the integrated absorption coefficient of the line (or the O.D. at a fixed frequency) has this same dependence. The possibility that

---

[298] M. S. Brodin and S. I. Pekar, *Zh. Eksperim. i Teor. Fiz.* **38**, 74 (1960); see *Soviet Phys. J.E.T.P.* (*English Transl.*) **11**, 55 (1960); *Zh. Eksperim. i Teor. Fiz.* **38**, 1910 (1960); see *Soviet Phys. J.E.T.P.* (*English Transl.*) **11**, 1286 (1960).

[299] I. S. Gorban' and V. B. Timofeev, *Dokl. Akad. Nauk S.S.S.R.* **140**, 793 (1961); see *Soviet Phys. "Doklady"* (*English Transl.*) **6**, 878 (1962).

interference effects are taking place has been considered and apparently ruled out in each case. Both crystals are, of course, anisotropic; this is a severe complication in anthracene but in Cu_2O it made possible a measurement on a quadrupole line of definite polarization. Thus in Cu_2O it is not easy to blame the effect on a slight spread of the polarization, since such a spread cannot excite a wave in a polarization perpendicular to the one under observation.

While these data are highly interesting and may establish qualitatively the anomalous wave effect, a complete theory has not been worked out and no quantitative comparison seems possible at present. The relative "initial intensities" I_+ and I_- must be computed; both waves are originally excited by the same photon beam. It is also important to note[298,300] that the effect depends for its observation on the coherence of the two waves over the entire thickness of the crystal. However, even though *exciton* mean free paths are often much smaller than the crystal thickness, the effect is not ruled out as an observable phenomenon. The photons exhibiting the Pekar effect will be those which in fact manage to traverse the whole crystal without being scattered.

The subject of anomalous waves is actually a part of the discipline of general physical optics, and we will not pursue it any further here. There has recently appeared a review and critique of the literature on the subject[301] in which it is emphasized that anomalous waves represent only one aspect of the general problem of spatial dispersion, the study of the effects on optical absorption of the **k** dependence of the dielectric function $\epsilon(\omega, \mathbf{k})$. Excitons represent only one type of excitation which can contribute to this **k** dependence. In the Cu_2O line studied by Gorban' and Timofeev, one already has "spatial dispersion" in the sense that the oscillator strength of the line is proportional to components of the photon wave vector. This area of microscopic physical optics will surely see more activity in the next few years.

---

[300] J. J. Hopfield, *Proc. Intern. Conf. Semicond. Phys., Prague, 1960* p. 443 (1961).

[301] V. L. Ginzburg, A. A. Rukhadze, and V. P. Silin, *Fiz. Tverd. Tela* **3**, 1835 (1961); see *Soviet Phys.—Solid State* **3**, 1337 (1961); see, also, S. I. Pekar, *Fiz. Tverd. Tela* **4**, 1301 (1962) [*English Transl.: Sov. Phys. Solid State* **4**, 953 (1962)]. It might be noted that the effect of anomalous waves on the Kramers-Kronig relations has been discussed by M. S. Brodin and A. F. Lubchenko [*Opt. and Spectr.* **7**, 48 (1959)].

IV. Transport Phenomena and Related Topics

12. Theory of Exciton Transport Phenomena

a. General[4,302,303]

Energy transport by excitons has been studied in a variety of ways ever since the early discussion of Frenkel[4] and Franck and Teller.[302] The latter were particularly concerned with the possibility of transport of energy culminating in storage of chemical energy, a process still of considerable interest to biochemists. Within the context of solid state physics, a simpler set of end products is expected, such as photons, heat (phonons), or free electrons and holes; the questions usually asked are (1) "does an exciton move (is **K** a good quantum number)?" and, more specifically, (2) "has it been directly verified that an exciton packet in Frenkel's sense can transport energy from one point to another in a crystal?"

Question (1) has been answered in the affirmative in the sense that certain observed energy levels depend in a predictable way on the wave vector of the state prepared by a photon. The most striking example of this is the Hopfield-Thomas $\mathbf{v} \times \mathbf{H}$ experiment in CdS (Section 6), but others are the magnetic-field intensity-interchange experiment (Section 9) and the very existence of a longitudinal-transverse splitting (Sections 3, 4), which imply participation of momentum states at other than $\mathbf{K} = 0$. In a weaker sense, the Zeeman experiments which prove that Wannier series are not impurity effects establish motion (if momentum conservation is assumed). The possibility of a different demonstration in the same general category is contained in Gross's suggestion[304] that a recoil energy (of $\sim 10^{-4}$ ev) might be observable in exciton emission. All of these facts bear on the

---

[302] J. Franck and E. Teller, *J. Chem. Phys.* **6**, 861 (1938).
[303] F. Seitz, *Rev. Mod. Phys.* **26**, 7 (1954).
[304] E. F. Gross, *Fiz. Tverd. Tela* **3**, 1899 (1961); see *Soviet Phys.—Solid State (English Transl.)* **3**, 1384 (1961).

initial state of excitation. Question (2) is much more complex in that it deals with the total history of the exciton. We discuss it now in terms of all the most likely channels for decay of the exciton.

(1) *Scattering* ($\nu \mathbf{K} \to \nu' \mathbf{K}'$). The exciton or polariton state prepared by absorption of light (or occurring as a by-product of impacts of ionizing particles or radiation) will have a fairly short lifetime against simple scattering into other exciton states. It is convenient, then, to think of "the exciton" as a packet which can be followed continuously from one scattering event to another. The effect of the scattering is to put the packet into states which may possibly be quite different from the original one. Since excitons normally comprise a low-density system ($\sim 10^{10}$ to 10^{12} cm^{-3}) they will thermalize to a Boltzmann distribution in roughly 10^{-13} to 10^{-11} sec unless some decay channel (below) competing with simple phonon scattering gives them a shorter lifetime. Their "thermal" velocities will be $(3kT/M^*)^{1/2} \approx 4 \times 10^6$ cm/sec, compared with their "initial" velocities $\hbar \varkappa / M^*$ ($\sim 5 \times 10^4$ cm/sec), where $\varkappa$ is the wave vector of the light in the medium. These numbers are appropriate only for direct ($\mathbf{K} \approx 0$) excitons. It should be noted that simple scattering may occur not only by phonons but also by point defects, dislocations, and surfaces. Scattering is discussed at greater length in Section 12b.

(2) *Radiative Decay.* An exciton packet may decay into a photon and some momentum- and energy-conserving phonon according to spontaneous transition probabilities which are essentially the same as the absorption probabilities (10.27). Toyozawa[244] has recently discussed lucidly the mechanisms of this process under several different circumstances. We note that this particular radiative decay involves moving excitons, as contrasted with an alternative in Section 12a (4). Radiative decay is handled in Section 13.

(3) *Dissociation.* A second fundamental channel for exciton decay is dissociation into an electron and hole pair, detectable as photoconductivity. Excitons in bound states do not carry current. Dissociation may presumably occur in any of several ways: phonon scattering,[121] collision ionization,[305] defect scattering, and field ionization (Section 6). In the first, the phonons and the kinetic energy of

---

[305] D. C. Northrup and O. Simpson, *Proc. Roy. Soc. (London)* **A244**, 377 (1958); S. I. Choi and S. A. Rice, *Phys. Rev. Letters* **8**, 410 (1962); *J. Chem. Phys.* **38**, 366 (1963).

the exciton must supply the energy needed for the dissociation, which is clearly $G/n^2 + \Delta E$ in the notation of Section 9, n being the hydrogenic quantum number. At high exciton densities two excitons may collide to produce an energetic electron-hole pair. Stationary defect scattering will be a useful mechanism when the exciton kinetic energy is originally large enough to supply the dissociation energy to a pair with small relative momentum. This can be confused experimentally in many cases, probably hopelessly, with excitation and subsequent dissociation of the defect itself. Clearly, too, in any dissociation process we may expect to be bothered by the possibility of "recombination radiation," confusable with direct radiative decay. Dissociation is discussed in terms of its principal macroscopic effect, "photoconductivity," Section 12c.

(4) *Self-Trapping.* This third fundamental channel results from large exciton-phonon coupling which is manifested in localization of the excitation at some definite site. The harmonic approximation to the electronic and vibrational states used in Section 10 must be replaced with the adiabatic approximation.[3,4,27] Thus, in complete analogy with luminescent impurity center theory, we expect at least three important secondary channels for decay: luminescence, occurring with a Stokes shift and therefore distinguishable from the preceding case (2) in most cases; radiationless decay with an accompanying burst of phonons (Seitz's "thermal spike"[303]); transfer of excitation by the dipolar, quadrupolar, or exchange mechanisms known in sensitized luminescence theory.[306] In this last case, there are at least two possible destinations for the transferred energy. It might jump to another site of the perfect lattice, in which case we have the "exciton diffusion" treated extensively by Trlifaj.[307] On the other hand, it might go to a defect (such as Tl^+ in an ionic crystal) at which any of the three processes could be repeated! This is one of the means by which we can realize "host-sensitized luminescence" (see, e.g., Teegarden[308]). Whether it is experimentally distinguishable from the alternative discussed immediately below is not at all certain.

---

[306] G. Cario and J. Franck, *Z. Physik* **17**, 202 (1923); T. Foerster, *Ann. Phys. (Paris)* [12] **2**, 55 (1948); D. L. Dexter, *J. Chem. Phys.* **21**, 836 (1953); see also C. C. Klick and J. H. Schulman, *Solid State Phys.* **5**, 97 (1957).

[307] M. Trlifaj, *Czech. J. Phys.* **6**, 533 (1956); **8**, 510 (1958).

[308] K. J. Teegarden, *U.S. Air Force Tech. Note* **TN58-917** (1958), unpublished.

(5) *Trapping.* If a defect has an appropriately large capture cross section, an exciton packet may conceivably become trapped there without (necessarily) having the benefit of strong phonon interactions. The assumption that this type of trapping process has a reasonable probability has motivated most research in which point defects are used as exciton "probes." If the exciton becomes trapped at the defect one has merely then prepared one of the excited states of the defect, whose known decay products can be observed. This, then, is another source of "host-sensitized luminescence." There is some, though very little, theoretical information available on the cross sections involved. Bound Wannier excitons[309] have recently played a part in the study of effective masses and other band parameters in semiconductors. In relating "bound excitons" to exciton transport questions their alternative description as excited defect states must be borne in mind explicitly; a photon might be exciting the defect directly. Trapping and self-trapping are discussed in Section 12d.

(6) *Miscellaneous.* Paranichev[310] has proposed an interesting decay which uses the magnetic field of an optical phonon to cause a singlet exciton to decay into a triplet. In molecular crystals where pure singlets and triplets may exist, the fields associated with optical modes must, however, be very small indeed. Pikus[311] and Korolyuk[312] have considered the effect of excitons on a macroscopic scale, namely a contribution to thermal conductivity, as discussed in Section 12e. Finally, it has been suggested that inhomogeneous electric[313] or magnetic[314] fields might predictably affect the diffusion of excitons. The author is not aware of any experiments in this area as yet.

b. Exciton Scattering

The essential quantity entering into any exciton scattering calculation will be the matrix element $\langle v'\mathbf{K}', a' | \mathcal{H}' | v\mathbf{K}, a \rangle$, where $v\mathbf{K}$

---

[309] M. A. Lampert, *Phys. Rev. Letters* **1**, 450 (1958); J. R. Haynes, *Phys. Rev. Letters* **4**, 361 (1960); D. G. Thomas and J. J. Hopfield, *Phys. Rev.* **128**, 2135 (1962).

[310] V. N. Paranichev, *Opt. Spectr. (U.S.S.R.) (English Transl.)* **9**, 62 (1960).

[311] G. E. Pikus, *Zh. Tekhn. Fiz.* **26**, 36 (1956); see *Soviet Phys.—Tech. Phys. (English Transl.)* **1**, 32 (1956).

[312] S. L. Korolyuk, *Soviet Phys.—Solid State (English Transl.)* **4**, 580 (1962).

[313] E. S. Gribnikov and E. I. Rashba, *Zh. Tekhn. Fiz.* **28**, 1948 (1958); A. Bierman, *Phys. Rev.* **130**, 2266 (1963).

[314] S. Nikitine and H. Haken, *Compt. Rend.* **250**, 697 (1960).

12. THEORY OF EXCITON TRANSPORT PHENOMENA

labels an initial exciton state, $\mathscr{H}'$ is a perturbation on the perfect crystal Hamiltonian $\mathscr{H}_0$, and $\nu'\mathbf{K}'$ is a final exciton state. The letters a, a' stand for the initial and final quantum numbers of any other relevant parts of the system to which the exciton is coupled, such as the phonon field or an impurity atom. As an example, the differential cross section in the Born approximation[315] per unit solid angle for intraband elastic scattering ($\nu = \nu'$) would be

$$\sigma_0(\mathbf{K}, \mathbf{K}') = \frac{2\pi\rho(\nu\mathbf{K}')}{\hbar I_0(\mathbf{K})} \overline{\overline{|\langle \nu\mathbf{K}', a | \mathscr{H}' | \nu\mathbf{K}, a \rangle|^2}}, \tag{12.1}$$

where the double bar indicates a thermal average over the states a, $\rho(\nu\mathbf{K}')$ is the density of (final) states at energy $E_{\nu\mathbf{K}'}$, and I_0 is the initial exciton flux. A total cross section is obtained by summing over all final states $\nu\mathbf{K}'$. As noted earlier, there are at least two effects which qualify any results obtained using simple exciton states $\Psi_{\nu\mathbf{K}}$ in computing the matrix element: (1) the "clothing" effect: the polarizing interactions which produce the effective dielectric constant (Section 7) also modify the detailed form of the exciton wave function. In practice this effect seems to enter mainly through renormalization of the effective charges (Section 7) and masses,[316] and we shall not treat it explicitly. The reader is referred to the papers of Haken[316] and Tulub[317] for details. Tulub points out a particular divergence introduced by using "bare" excitons; the mean free path for exciton-optical phonon scattering becomes infinite in the case $m_e^* = m_h^*$, if the simple perturbation results of Section 10 are used. He computes a finite result using Low's polaron method.[318] (2) The "polariton" effect: when the pure exciton state $\Psi_{\nu\mathbf{K}}$ has nearly the same energy, wave vector, and polarization as the radiation field, the polariton states (9.43) are probably the states whose dynamics are of most relevance to exciton decay theories. The cross section (12.1) holds for polariton-polariton scattering provided that we include a factor

$$| c_2(\nu\mathbf{K}')^* c_2(\nu\mathbf{K}) |^2$$

---

[315] See, e.g., L. I. Schiff, "Quantum Mechanics," 2nd ed., Sects. 39-40. McGraw-Hill, New York, 1955.

[316] H. Haken, *Z. Physik* **155**, 223 (1959).

[317] A. V. Tulub, *Zh. Eksperim. i Teor. Fiz.* **36**, 1859 (1959); see *Soviet Phys. J.E.T.P.* *(English Transl.)* **9**, 1325 (1959).

[318] F. E. Low, *Phys. Rev.* **97**, 1392 (1955).

[cf. (9.43)] to account for the fraction of pure exciton mixed into the final and initial states, and use the initial polariton group velocity in computing the incident intensity $I_0(\mathbf{K})$. These considerations appear to have been brought into practice only in some of the Russian literature.[265]

(1) *By Phonons.* In the case of phonons, the perturbation consists of a field of lattice displacements. Details of this scattering process are therefore contained in Section 10. Generally exciton scattering by phonons is characterized by a mean free path, which may be computed from $\tau_{\nu\mathbf{K}} v_{\nu\mathbf{K}}$, where $v_{\nu\mathbf{K}}$ is the group velocity of the exciton packet centered (in momentum space) at $\mathbf{K}$. The mean free path as limited by acoustic phonons may thus be computed using the lifetime discussed in Section 10 [Eq. (10.28)], under the same assumptions of high temperature and long wavelength of the exciton. In this case the dependence of the mean free path on temperature (T), deformation potential difference ($E_n - E_m$), mass of the atoms (M), velocity of sound (u), and reduced mass of the exciton (μ) is

$$l \propto Mu^2/\mu^2(E_n - E_m)^2 \, kT.$$

In the case of optical phonon scattering, this dependence is

$$l \propto (M/\mu) \, e^{\hbar\omega_0/kT}$$

where ω_0 is the optical phonon frequency. Remarkably few attempts have been made to apply these results for mean free paths to real crystals, probably again because of a lack of knowledge of the necessary parameters. In the Frenkel case, Agranovich and Konobeev[247] obtain an estimated path of 20 A for a thermal exciton in a simple model of anthracene.

Other estimates[254,262] in ionic and rare gas crystals seem to indicate mean free paths of this same order of magnitude at 300° K and 20° K, respectively. One can probably expect mean free paths of perhaps an order of magnitude larger in the alkali halides at low temperatures, say 100 or 200 A.

(2) *By Point Defects.* The theory of simple (elastic) collisions with points defects has been considered by Kachlivskii[319] and

[319] Z. S. Kachlivskii, *Soviet Phys.—Solid State (English Transl.)* 3, 361 (1961) (F centers); 3, 1554 (1962) (impurity centers); 3, 2091 (1962) (U centers).

Selivanenko.[320] Exactly as in other areas of transport theory, impurity scattering is held to be responsible for limiting the exciton mean free path at low temperatures, where phonon scattering becomes small. However, Kachlivskii estimates that a thermal Wannier exciton in an alkali halide will have a mean free path of 10^4 Å in the presence of a density of 10^{19} impurities or U centers per cm³, and that point defect scattering will predominate over lattice scattering below about 10° K in this case. The "lattice scattering" over which the point scattering predominates below 10° K appears to have been taken as optical-mode scattering only. A mean free path of 10^4 Å is much larger than one expects; this result illustrates the general ineffectiveness of simple elastic point-defect scattering. Inelastic interactions with point defects and surfaces are undoubtedly the true source of residual scattering when lattice scattering becomes negligible.

(3) By Dislocations. Hrivnak[321] computes the mean free path of Wannier excitons in the presence of edge dislocations, using an electrostatic strain-field potential.[322] He presents estimates of dislocation densities above which dislocation scattering will predominate over phonon scattering at low temperatures. Because of the large number of parameters involved for a given crystal, the reader is referred to his paper for details. It may be mentioned that his mean free path becomes infinite when $m_e^* = m_h^*$, probably because of the separated wave function used in the calculation.[317] The effect of dislocations on absorption edge shapes has been considered by Blakney and Dexter[323] in the context of the one-electron picture; this theory probably could be profitably extended to account for exciton states.

(4) By Surfaces. Since very little has been done in the area of exciton-surface interaction in the dynamical sense, we take this opportunity to point out the existence of theoretical work[324,325]

---

[320] A. S. Selivanenko, *Soviet Phys.—Solid State (English Transl.)* **3**, 733 (1961) (molecular crystals).

[321] L. Hrivnak, *Czech. J. Phys.* **7**, 390 (1957).

[322] See, e.g., D. L. Dexter and F. Seitz, *Phys. Rev.* **86**, 964 (1952).

[323] R. M. Blakney and D. L. Dexter, *Proc. Phys. Soc. (London)* (1955): *Rept. Bristol Conf. on Defects in Crystalline Solids, 1954*, p. 108 (1955).

[324] A. S. Selivanenko, *Zh. Eksperim. i Teor. Fiz.* **32**, 75 (1957); see *Soviet Phys. J.E.T.P. (English Transl.)* **5**, 79 (1957).

[325] R. E. Merrifield, *J. Chem. Phys.* **28**, 647 (1958).

on "surface excitons"; Katul and Zahlan[326] believe they have observed surface excitons in tetracene. One may generally expect that surfaces play a large role in destroying the coherence of a photon and any excitations it produces, especially when the absorption coefficient of a material is so high that most absorption takes place near the surface. In his calculations on radiative lifetimes, Toyozawa[244] has assumed specular reflection of excitons by surfaces.

c. Photoconductivity and Photoemission

It is the purpose of this section to consider only the overlap between studies of excitons and photoconductivity. The latter is a vast subject in solid state physics, and a compendium of recent work may be found in the proceedings of the 1961 conference at Cornell.[327] Since the exciton is electrically neutral, it carries no current, as may be verified by computing the expectation value of the current operator in an exciton state:

$$\int \Psi^*_{\nu K} \mathbf{J} \, \Psi_{\nu K} \, d\tau = 0. \tag{12.2}$$

This has been done by Wannier[5] and, for the Frenkel exciton, by Seitz.[328] In a real crystal, however, one usually *does* observe photoconduction as a result of optical absorption in the exciton band; this clearly requires that some dissociation mechanism be operative. As representative examples of this phenomenon, we mention the photoconductivity resulting from exciton-band absorption in Cu_2O,[329,330] HgI_2,[331] CdS[332] and organic crystals,[333] and the photoemission under similar circumstances in BaO[334,335] and alkali halides.[336,337] We now

[326] J. Katul and A. B. Zahlan, *Phys. Rev. Letters* **6**, 101 (1961).
[327] Published as vol. 22 of *Phys. Chem. Solids* (1961).
[328] F. Seitz, "The Modern Theory of Solids," p. 417. McGraw-Hill, New York, 1940.
[329] V. P. Zhuze and S. M. Ryvkin, *Dokl. Akad. Nauk S.S.S.R.* **77**, 241 (1951).
[330] J. H. Apfel and A. M. Portis, *Phys. Chem. Solids* **15**, 33 (1960).
[331] E. F. Gross, A. A. Kaplianskii and B. V. Novikov, *Zh. Tekhn. Fiz.* **26**, 697 (1956).
[332] E. F. Gross and B. V. Novikov, *Phys. Chem. Solids* **22**, 87 (1962).
[333] See, e.g., the papers collected in "Symposium on Electrical Conductivity in Organic Solids." (H. Kallman and M. Silver, eds.). Wiley (Interscience), New York, 1961.
[334] H. R. Philipp, *Phys. Rev.* **107**, 687 (1957); R. J. Zollweg, *ibid.* **111**, 113 (1958).
[335] E. Taft, H. Philipp, and L. Apker, *Phys. Rev.* **113**, 156 (1959).
[336] L. Apker and E. Taft, *Phys. Rev.* **81**, 678 (1951).
[337] H. R. Philipp and E. A. Taft, *Phys. Rev.* **106**, 671 (1957); E. A. Taft and H. R. Philipp, *Phys. Chem. Solids* **3**, 1 (1957).

discuss these phenomena in terms of the possible dissociation mechanisms.

(1) *Dissociation by Phonons.* In a pure crystal, phonons are the principal perturbation required for the breakup of an exciton. The only quantitative work on this mechanism seems to have been done by Lipnik[338] and Goodman and Oen.[121] Lipnik treats the Wannier exciton in nonpolar crystals and obtains the following lifetime against dissociation by single-phonon processes:

$$\tau_d = \frac{(2\pi\hbar^2)^{3/2}}{3^{1/2}\mu(kT)^2}\frac{1}{\sigma_{ph}}e^{\Delta/kT} \tag{12.3}$$

where μ is the exciton reduced mass, k is Boltzmann's constant, $\Delta = G/n^2$ is the binding energy of the exciton under consideration, and σ_{ph} is a cross section for phonon-induced recombination or decay of the exciton, which Lipnik finds to be relatively temperature-independent. In Cu_2O, the lifetime of a thermal $1s$ exciton computed[338] from Eq. (12.3) is 6×10^{-12} sec at room temperature and 6×10^{-6} sec at $100°$ K, using $\Delta = 0.15$ ev. The cross section σ_{ph} is found to be quite large in both Cu_2O and Ge, being of the order of 10^{-13} cm^2. Lipnik notes the contrast between this problem and the case of capture of an electron by an impurity center,[339] pointing out that when recombination or dissociation take place here, the kinetic energy of the exciton is available to conserve energy and multiphonon processes are not *necessarily* important. Lipnik finds that the calculations of cross sections are quite sensitive to the form of the conducting states chosen; in particular, one obtains quite different results when using plane waves rather than hydrogenic continuum states, the latter being presumably a better choice.

Goodman and Oen's calculation[121] is an interpretation of Philipp's BaO data[334]; although the formalism (based on an exciton comprised of clothed particles) and crystal (ionic) are vastly different from those of Lipnik, the orders of magnitude of the quantities involved are very similar. The observed activation energy Δ is 0.15 ev, and the estimated lifetime is $\tau_d \gtrsim 10^{-11}$ sec. Goodman and Oen's theory agrees with experiment, however, only if it is assumed that at least two phonons are involved in the dissociation.

---

[338] A. A. Lipnik, *Fiz. Tverd. Tela* **3**, 2322 (1961); see *Soviet Phys.—Solid State (English Transl.)* **3**, 1683 (1962), and references therein.
[339] See, e.g., M. Lax, *Phys. Chem. Solids* **8**, 66 (1959).

The above work has been done on models in which the exciton under consideration is at an energy lower than all available conduction states. Naturally, excitons of greater energy than the smallest band gap are much more readily ionized by any kind of imperfection. We shall not treat this case, examples of which might include the direct excitons in Ge and Si (see Fig. 29) and the second direct exciton in the iodides, which appears to lie above the lowest band gap.

(2) Collision ionization.[305] A crystal containing two excitons is in a state whose energy is the same as that of a free hole-electron pair of high kinetic energy. Choi and Rice[305] have shown that transitions from the former to the latter cannot be neglected in cases of practical interest.[340] Using essentially the one-electron Frenkel model (Section 3), they find a dissociation rate of

$$dn/dt \approx - (n^2/Na^6) \times 10^{17} \text{ sec}^{-1} \qquad (12.4)$$

in a tetragonal crystal in which $c \gg a$ (a and c are the unit cell dimensions), where n is the density of randomly distributed excitons, and N is the number of cells in the crystal. It would be of considerable value to refine this theory to account for nonrandom exciton distributions (in particular, distributions produced by high-intensity coherent light beams), in which case the dissociation rate might be considerably higher. The dependence of Eq. (12.4) on n^2 means that one expects a similar dependence on the intensity of the light source producing the excitons, but its experimental verification is made difficult by generation of carriers at surfaces. Moreover, when electron-hole recombination probabilities are high, the observed photocurrents become linear in the light intensity, as pointed out by Choi and Rice.[305] Evidence for the quadratic dependence in anthracene has been offered by Silver *et al.*[340]

(3) *Dissociation by, or Ionization of, Static Imperfections.* A static imperfection cannot readily dissociate an exciton which lies below all conducting states, merely on grounds of energy conservation. Use of the exciton's kinetic energy would be required to overcome its binding energy, and a photoproduced exciton has little kinetic

---

[340] E.g., R. G. Kepler, *Phys. Rev.* **119**, 1226 (1960); M. Silver, D. Olness, M. Swicord, and R. C. Jarnagin, *Phys. Rev. Letters* **10**, 12 (1963). Kepler et al. [R. G. Kepler, J. C. Caris, P. Avakian, and E. Abramson, *Phys. Rev. Letters* **10**, 400 (1963)] appear to have observed triplet-triplet exciton collisions, inferred by fluorescence attributed to one of the by-products, singlet excitons.

energy. An *inelastic* dissociation process would be rather rare, since it would require that the exciton encounter an excited defect which could give its excitation energy to the exciton. The case of dissociation resulting from collisions with neutral impurities has been considered by Lipnik,[338] who concludes quantitatively that for most impurity densities of interest phonon dissociation will be the dominant mechanism.

We now turn to *ionization* of imperfections by excitons, a process in which an exciton disappears, giving its entire excitation energy to the freed electron or hole. It is this process which was brought into prominence by the photoemission experiments of Apker and Taft[336,341] on the alkali halides. Apker and Taft concluded, on the basis of the spectral photoemissive response in evaporated films containing large numbers of F centers, that energy was being transferred from excitons to the F centers; Philipp and Taft[337] corroborated these results using cleaved single crystals and concluded that the effective cross section for exciton-induced ionization is less than 10^{-14} cm^2.

Ionization of F centers by excitons has been studied theoretically by Dexter and Heller,[342] Toyozawa,[163] and Trlifaj.[343] Dexter and Heller treated the process in terms of a two-center Auger (autoionization) transition, concluding that, in the alkali halides, the probability of such a transition is 10^{12} sec^{-1} for a nearest neighbor Auger transition. These authors also treat the problem of F-center creation from an exciton and a vacancy. Toyozawa[163] estimates that at least 0.5×10^{16} F centers per cm^3 are necessary to make the ionization process more probable than radiative emission. Toyozawa's calculation is based on the electronic polaron formalism discussed in Section 7, and it is interesting to note that in his model transverse excitons do not interact with the F-center electron; thus an optically produced exciton must be assumed to thermalize into a longitudinal state before ionizing the F center. Trlifaj[343] has computed cross sections for the following end products of an exciton colliding with an F center: F' center plus free hole; free electron plus an electron bound to the vacancy. To the author's knowledge, no numerical applications of his formalism are available.

Although there seems to be a need for it, little effort seems to

---

[341] M. H. Hebb, *Phys. Rev.* **81**, 702 (1951).
[342] D. L. Dexter and W. R. Heller, *Phys. Rev.* **84**, 377 (1951).
[343] M. Trlifaj, *Czech. J. Phys.* **9**, 446 (1959).

have been made in the literature to extend these calculations to encompass ionization of other donor centers by the various types of excitons. There has, of course, been a lot of qualitative discussion of the process, especially in connection with energy transport questions. This is discussed in Section 12f.

(4) *General Remarks.* It would appear hopeless, from this multitude of dissociation processes, to try to verify the zero-current theorem [Eq. (12.2)] for an exciton state, but it is at least reasonable to verify that the photocurrent is very small in a pure crystal with a fairly large exciton binding energy. The alkali halides have thus been studied for many years; Ferguson[344] was the first to tentatively verify the contention, and an extensive investigation was carried out by Taylor and Hartman.[345] However, these authors saw no photocurrent anywhere in the spectral region covered, which means that no statement could be made about the production of carriers in exciton absorption relative to that in band-to-band absorption. Teegarden[346] and Nakai[347] finally verified that photoconductivity does set in at an energy corresponding to the band gaps in KI and RbI, and that the currents caused by absorption in the lowest exciton band are at least a factor of 30 smaller even at room temperature. Similar results have been obtained by Kuwabara and Aoyagi[348] in KI, KBr, and KCl. The band gaps mentioned immediately above are those deduced from the position of the shoulder appearing between the lowest two exciton peaks in the iodide absorption spectrum,[109] and which also is featured in the iodides' photoemission spectra.[337]

In concluding this brief subsection we would like to mention two recent theoretical observations bearing on photoconductivity questions. The first is the proposition by Hopfield[267] that, whenever photoconductivity results from exciton dissociation, the photoconductive response will be a function only of the spatial distribution, hence of absorption coefficient, in that spectral region. This is based on the assumption that in a given absorption band, the light produced roughly the same final states, regardless of the energy of the photon as it ranges over the band. The absorption shape itself is here held

---

[344] J. N. Ferguson, *Phys. Rev.* **66**, 220 (1944).
[345] J. W. Taylor and P. L. Hartman, *Phys. Rev.* **113**, 1421 (1959).
[346] K. J. Teegarden, *U.S. Air Force Tech. Note* **TN59-303** (1959), unpublished.
[347] Y. Nakai and K. J. Teegarden, *Phys. Chem. Solids* **22**, 327 (1962).
[348] G. Kuwabara and K. Aoyagi, *Phys. Chem. Solids* **22**, 333 (1962).

to be a consequence of the distribution of available intermediate states for the process. Hopfield discusses the range of applicability and methods of verifying this conjecture experimentally.

The second observation is that of Dykman and Pekar,[349] who, using anomalous wave theory (Section 11), conclude that certain light waves can have very high amplitudes in regions of small index of refraction such as the vicinity of an absorption peak; thus, at these frequencies an impurity might be much more readily ionized. This remarkable suggestion is subject to several qualifications (given by Dykman and Pekar), and to some of the objections to anomalous wave theory itself (Section 11). The authors note that the effect does not involve excitons in any direct way, and the result seems to be that we have another imponderable to consider when analyzing transport experiments.

d. Trapping and Diffusion

Section 12a (4) and (5) form the introduction to this subsection. Here, we discuss somewhat more fully the ideas of self-trapping, diffusion, and trapping by defects.

(1) *Self-Trapping.* Peierls's and Frenkel's criteria for "absorbers" and "self-trapping" were similar, and amounted to a requirement that the exciton-phonon coupling constant G (Section 10) be larger than the matrix element of $\mathcal{H}_0$ connecting two states of excitation localized on nearest neighbors. This matrix element was called $E_{ii'}$ in Section 3 in a discussion of the Frenkel exciton,[42] and in general it has the order of magnitude of the exciton bandwidth. Translated into lifetimes: an exciton may become self-trapped if $\tau_l \sim \hbar/G$ is much shorter than $\tau_j \sim \hbar/E_{ii'}$, the jump time between sites. Here τ_l is roughly the time taken for the lattice to relax around a localized excitation. Large exciton-phonon coupling constants mean short mean free paths when they occur in considerations of nonlocalized exciton states; we may say qualitatively that a mean free path of the order of a lattice constant is a strong indication that self-trapping will take place.

Self-trapping may always be called upon to explain the complete disappearance of exciton energies in a crystal. In this case it is assumed

---

[349] I. M. Dykman and S. I. Pekar, *Zh. Eksperim. i Teor. Fiz.* **37**, 510 (1959); see *Soviet Phys. J.E.T.P. (English Transl.)* **10**, 361 (1960).

that a localized excited state is prepared which decays completely by emission of a large number of localized phonons,[350] exactly as in the theory of radiationless decay of impurity excitations, and like a thunderclap in the middle of the desert, nothing is there to "hear" it. Seitz[303] made considerable use of the thermal spike concept in discussing the kinetics of color center formation and diffusion, and Ueta and Hirai[351] have recently attributed R-center bleaching to these thermal spikes. A detailed theory of the process of self-trapping has not been developed; such a theory would be invaluable in assessing the valid ranges of two popular alternative assumptions; that the optically produced exciton thermalizes immediately, or is trapped immediately. It is to be noted, however, that one rather interesting prediction was made in an old paper by Dykman and Pekar[352]: in an ionic crystal, self-trapping of a Wannier exciton as a result of optical phonon-exciton interactions is likely if m_h^*/m_e^* is either greater than 10 or less than 1/10.

(2) *Diffusion.* If self-trapped, the exciton may get rid of its energy by emission (Section 13) or by resonant transfer to another site. The latter process corresponds precisely to the sensitized luminescence process,[306] which is ordinarily regarded as a mechanism for impurity-to-impurity energy transfer. It has been applied to excitons by Trlifaj,[307] whose results are in fairly good quantitative agreement with experimental values of the diffusion length for excitation in anthracene as measured by Simpson.[353]

(3) *Trapping by Defects.* We should like to emphasize further that an "exciton trapped at a defect" is just another name for "excited state of a defect," and that this is perhaps more than just an excuse to avoid a lengthy discussion of the electronic structure of trapped excitons. In many cases,[309,354] the trapped exciton or "bound exciton complex" is studied for the main purpose of determining band parameters; but in others, the defect is used as a probe to catch the exciton and it may be claimed that some specific exciton motion has taken

---

[350] This process, as applied to excitons, has been discussed quantitatively by H. Stumpf, *Z. Naturforsch.* **14a**, 659 (1959).

[351] M. Ueta and M. Hirai, *J. Phys. Soc. Japan* **14**, 546 (1959).

[352] I. M. Dykman and S. I. Pekar, *Dokl. Akad. Nauk S.S.S.R.* **83**, 825 (1952).

[353] O. Simpson, *Proc. Roy. Soc.* **A238**, 402 (1956).

[354] D. G. Thomas and J. J. Hopfield, *Phys. Rev. Letters* **7**, 316 (1961), and references therein.

place. It is here that the distinction becomes more than semantic. For if the excited state of the defect could have been prepared by a photon directly, the probe has lost its usefulness.

Certain bound excitons have a long-standing relationship to exciton theory and deserve special mention. These are the α- and β-bands,[355] absorption bands near the fundamental edge due to excitons "bound" to negative ion vacancies and F centers, respectively. A calculation based on the simple electron transfer model[355] gives reasonable values for the positions of the α- and β-bands. The α-band computations have recently been extended to include transfer of the electron to other than nearest neighbors,[356] and the β-band calculations have been refined[357] to account for the point symmetry and spin-orbit interaction of the halogen ions, in analogy with Overhauser's calculation. Although Fuchs predicts an observable spin-orbit splitting in the β-band, none seems to be observed.[358] Another example of the "bound exciton" is the absorption due to substitutional halide ion impurities in the alkali halides. Wherever the impurity halide has a smaller electron affinity than that of the host halide, a new absorption band is revealed, and the Hilsch-Pohl rule for intrinsic absorption may be used (with the impurity's electron affinity and the host's lattice constant) with uncanny success in predicting this band's location.[6] Finally, we note that any of the excited states of a Tl^+ ion in an alkali halide can be considered a "bound exciton."

e. Contribution to Thermal Conductivity

Normally, in insulating crystals and semiconductors heat is transported only by phonons. However, in small-band-gap semiconductors, exciton states have sufficiently low excitation energies that they may become appreciably populated at ordinary (and high) temperatures and contribute to heat flow (without, of course, any associated electrical current). Herring[359] points out that the contribution of excitons to thermal conductivity will be swamped by that of electrons and holes whenever the exciton excitation energy is of the same order as the gap

---

[355] F. Bassani and N. Inchauspé, *Phys. Rev.* **105**, 819 (1957).

[356] A. A. Tsertsvadze, *Fiz. Tverd. Tela* **3**, 336 (1961); see *Soviet Phys.—Solid State (English Transl.)* **3**, 241 (1961).

[357] R. Fuchs, *Phys. Rev.* **111**, 387 (1958).

[358] C. J. Delbecq, P. Pringsheim, and P. H. Yuster, *J. Chem. Phys.* **19**, 574 (1951); **20**, 746 (1952).

[359] C. Herring, *Phys. Chem. Solids* **8**, 543 (1959).

(i.e., $G \ll E_G$), because of the considerably larger portion of phase space available to the pair (continuum) states. Herring suggests that excitons will make an observable contribution only when $G \gtrsim \frac{1}{2} E_G$, which, as we have seen, is rarely the case. Nonetheless, calculations have been made[311,312] using standard transport theories in an attempt to explain thermal conductivity data on InSb[360] and PbTe.[361] Korolyuk[312] concludes that the InSb data can be explained on the basis of participation of excitons of excitation energy 0.26 ev (which is larger than the 0.24 ev band gap!), provided that it is assumed that their mean free paths are determined mainly by phonon scattering rather than dissociation events. Since this seems difficult to justify, the importance of excitons to thermal conduction remains considerably in doubt.

f. The Energy Transfer Question

(1) *General.* Let us distinguish three kinds of exciton transport of energy. The first will be called merely *electromagnetic* transport, and is defined as the transport of energy by a simple polariton state having a large photon component. Such a state is created, as discussed in Section 9, when an external photon enetrs a crystal. It contributes one quantum of excitation to the electromagnetic flux $nc\,\mathsf{E}^2/4\pi$, and, in the absence of scattering, will travel to the other side of the crystal. As discussed in Section 6, the exciton component of these waves has been identified as a momentum state by the magnetostark effect. The second kind of energy transport is called "hopping" transport, defined in terms of a self-trapping of energy followed by subsequent jumps, as in sensitized luminescence, and as discussed by Trlifaj. The third kind will be called *wave-packet* transport, which encompasses all *other* transport of energy attributable to excitons. Wave packet transport occurs between the time that the initial polariton becomes scattered into nonradiative states (mostly exciton-like) and when it decays into some other form (a free pair; a photon; phonons; or a self-trapped exciton). Most workers who have tried to demonstrate energy transport by excitons have not usually made a distinction such as this among types of energy transfer, but they seem to have implicitly hoped to verify *wave-packet* transport. One can hardly deny the

[360] G. Busch and M. Schneider, *Physica* **20**, 1084 (1954).
[361] E. D. Devyatkova, *Zh. Tekhn. Fiz.* **27**, 461 (1958); see *Soviet Phys.—Tech. Phys.* *(English Transl.)* **2**, 414 (1958).

existence of electromagnetic transport, especially since it is responsible for any of the transmitted light which does emerge from an absorbing crystal. Furthermore, although one cannot ignore the theoretical and practical interest attached to *hopping* transport, wavepacket transport is much closer to the original spirit of Frenkel's proposed excitation packet concepts, and is worth searching for and comparing with the hopping process, if possible.

For further clarification of the three kinds of energy transport, we refer to Fig. 34 and note that as a beam of polaritons sets out in the

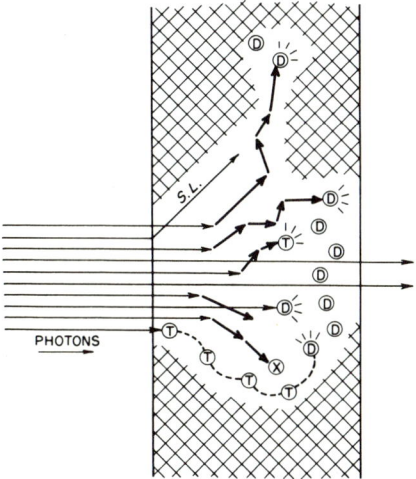

FIG. 34. Illustrating various modes of exciton transport of energy. The thin lines represent unscattered polaritons (i.e., crystal photons) and the heavy arrows represent exciton wave packets. Dashed lines illustrate "hopping" transport, by means of the mechanism of sensitized luminescence. Code: D, detector; T, trapped exciton; X, impurity; S.L., scattered light.

crystal, scattering peels off many of them and they are turned into exciton wave packets (heavy arrows). A few of the original excitations get all the way through and transport energy "electromagnetically." Some of the wave packets may become trapped excitons.

Ideally one wishes to place detectors in the crystal which present a sizeable cross section for capture of the exciton packets and which decay into unambiguously identifiable products. To avoid detection of the electromagnetic component of the energy flux, the detectors should ideally have no cross section for photon absorption; since this is impractical, they might be placed at the far end of the crystal where little electromagnetic flux is present. The exciton transport question might be phrased "can an exciton go where a photon cannot?"

We now briefly survey the present status of exciton transport experiments and attempt to fit them into the above framework.

(2) *Alkali Halides.* The early experiments of Apker and Taft[336] indicated energy transfer from excitons to F centers (the detectors). Photoemission was observed as a result of F-center ionization, and it is generally believed[337] that energy transfer takes place at least over short distances. The transfer can best be regarded as of the hopping type if Dexter and Heller's Auger mechanism[342] is operative.

Ueta and Tomura and colleagues[362–366] have recently carried out considerable research on the alkali halides using a variety of point-defect detectors such as Tl^+ ions, foreign halides, and color centers. For example, using U centers as detectors, Ueta and co-workers[365] find energy transport over a distance of 300 A in KBr. Results such as this one have been deduced with a great deal of effort from the wavelength-, temperature-, and detector-concentration-dependence of luminescence and photocurrents. A critical review of the large number of assumptions which have been necessary in this analysis would consume all too many pages of this review. We should however like to make the following remarks which may be of use to those desiring to extrapolate the alkali halide results or to plan similar experiments.[367] The fundamental absorption edges of the alkali halides are extremely sensitive to impurity effects, and it follows that light absorbed in the sensitive wavelength region will preferentially excite the impurities and their immediate environs. Examples of such impurities are substitutional Br^- in KCl, I^- in KBr, and Tl^+ in any alkali halide. The luminescence observed by Tomura and Kaifu and attributed to exciton energy transfers to Tl^+ ions is characteristic of excitation into

[362] Early work is reviewed by Ueta, ref. 17.

[363] M. Tomura and Y. Kaifu, *J. Phys. Soc. Japan* **15**, 314 (1960) (KBr: I).

[364] M. Tomura and Y. Kaifu, *J. Phys. Soc. Japan* **15**, 1295 (1960); M. Tomura, *ibid.* **15**, 1508 (1960) (KI:Tl).

[365] M. Ueta, M. Hirai, and H. Watanabe, *J. Phys. Soc. Japan* **15**, 593 (1960) (U → F center conversion).

[366] M. Ueta, M. Hirai, and M. Ikezawa, *J. Phys. Soc. Japan* **16**, 528 (1961) (photoconductivity from U centers).

[367] The author is indebted to Professor K. J. Teegarden for enlightening discussions on these points.

the "D" band[368] of that impurity itself[369]; the D band is located quite near the fundamental edge of the crystal and may well be intercepting photons directly (i.e., "detecting electromagnetic transfer") in Tomura's experiment. In Ueta's experiments on U- to F-center conversion and on photoconductivity induced by U-centers ionization by mobile excitons, there seems to be abundant evidence that these processes are sensitive to, and could be even entirely caused by, the presence of halide impurities which absorb in the fundamental absorption edge (Br^- in KCl; I^- in KBr). Also not ruled out in Ueta's experiments is the possibility that excitons are immediately trapped, and the detectors receive their energy by one "hop" of the resonance-transfer type. Our contention is that an approach based strictly on the process of elimination (of alternative mechanisms) must be used in the analysis of exciton energy transport experiments, and that "elimination" has not proceeded far enough in the case of the alkali halides.

(3) *Semiconductors.* Because of the relatively weak exciton-lattice interaction in most semiconductors, as indicated by narrow line widths, self-trapping is probably fairly unlikely and one needs discuss only wave packet transport. Long-range photoconductive effects in CdS were at one time believed to occur as a result of exciton migration. For detection purposes, photocurrents were sought in a region of the crystal not illuminated by the incoming photon beam. It appears to have been demonstrated, however, that scattered light was responsible for the observed current excitation.[370] A discussion of these experiments is given by Allen,[371] who applies the various scattering theories described earlier to GaAs and CdS, showing that exciton diffusion lengths are at least three orders of magnitude smaller than the lengths required to explain the experiments. Exciton wave-packet transport therefore seems not to have been verified in semiconductors.

(4) *Organic Crystals.* Hopping transport or "diffusion" of exciton

---

[368] See P. H. Yuster and C. J. Delbecq, *J. Chem. Phys.* **21**, 892 (1953) (KI:Tl); a similar band is found in KCl:Tl [see the curves of J. E. Eby and K. J. Teegarden quoted in R. S. Knox, *Phys. Rev.* **115**, 1095 (1959)].

[369] Luminescence as a result of D-band absorption is in most respects identical to that observed by M. Tomura (K. J. Teegarden and R. F. Edgerton, unpublished).

[370] I. Broser and R. Broser-Warminsky, *Phys. Chem. Solids* **8**, 177 (1959).

[371] J. W. Allen, *Proc. Intern. Conf. Semicond. Phys., Prague, 1960* p. 435 (1961).

energy has long been known in organic solids.[353,372] Simpson[353] determined a diffusion length of 460 A in anthracene by the use of naphthacene detector molecules. There is considerable current activity in this field; of particular interest is the discovery[373] that triplet exciton diffusion occurs as rapidly as singlet exciton diffusion in naphthalene. This effect may arise from relatively large intermolecular matrix elements in the triplet state, as noted by Nieman and Robinson[374]; these matrix elements are also known to be relatively large in solid argon (Section 3), where rapid triplet energy transfer would also certainly be of interest were it not that intermediate coupling prevents states of pure spin multiplicity from being of experimental interest.

13. Radiative Decay

Radiative decay is treated separately from other exciton decay processes not because a large body of theory exists but because it appears likely that its importance will increase rapidly in the near future, as a result of recent intensive studies of stimulation of emission of radiation in the optical region of the spectrum. Two distinct processes may be studied, as implied in Section 12: emission from trapped excitons and from "free" excitons.

For many years the possibility of exciton emission was largely discounted, mainly on the grounds that no emission was found in the alkali halides, where most early experiments were performed. Emission is now observed from Frenkel excitons in the organics, from Wannier excitons in semiconductors, and even, possibly, from trapped excitons in the alkali halides themselves.

a. Emission from "Free" Excitons[244,375,376]

By definition, free-exciton luminescence involves initial electronic states which may be very similar, or close in energy to, the states responsible for absorption. Thus one might look for emission in the

---

[372] H. C. Wolf, *Z. Physik* **139**, 318 (1954); see also the reviews by McClure[14] and Wolf.[15]

[373] M. A. El-Sayed, M.T. Wauk, and G.W. Robinson, *Molecular Phys.* **5**, 205 (1962).

[374] G. C. Nieman and G. W. Robinson, *J. Chem. Phys.* **37**, 2150 (1962).

[375] F. A. Kroger and H. J. G. Meijer, *Physica* **20**, 1149 (1954).

[376] Yu. M. Popov and A. S. Selivanenko, *Opt. Spectr. (U.S.S.R.) (English Transl.)* **9**, 135 (1960).

13. RADIATIVE DECAY

vicinity of the absorption edge or "edge emission." Direct and indirect transitions may be studied separately. In the direct case, the complication introduced by strong exciton-photon coupling must be considered explicitly, and this has been done in a unique calculation by Toyozawa.[244] He considers excitons interacting linearly with acoustic phonons, the excitons initially thermally distributed fairly high on the curve T_1 (Fig. 32). For emission to occur, the excitons must work their way back toward the steep part of curve T_1, where their photon component will be large enough to stimulate an external photon. To get there, they must pass through the inflection region of curve T_1, which Toyozawa calls a "bottleneck" because, as he shows, the transit time through that region in a typical model crystal is about 10^{-9} sec, which is 3 orders of magnitude longer than typical phonon-scattering lifetimes. The reasons for this slow leakage back to radiative states are low densities of exciton states and weak exciton-phonon coupling at small wave vectors (as shown in Section 10). Although Toyozawa does not consider the case in which the excitons are initially created below the bottleneck, it seems clear that the bottleneck can operate in both directions. Thus excitations produced by fundamental absorption-tail irradiation may never have a chance to thermalize, as it is often assumed they do.

There seems to have been no quantitative work done on the case of optical-phonon-assisted exciton decay, but this mechanism was introduced by Kroger and Meijer[375] in the interpretation of CdS edge emission. The particular "green" edge emission which interested Kroger and Meijer has been found to be due to an entirely different recombination process, rather than exciton decay.[377] On the other hand, "blue" emission which is coincident in energy with CdS exciton absorption lines[87] undoubtedly corresponds to free-exciton decay, and other semiconductors are currently under investigation.[378] Gross[304] proposes a search for exciton recoil energies by accurate determination of absorption and emission wavelengths. Emission from excitons in organic crystals is discussed in other recent reviews.[379]

[377] See J. Lambe, C. C. Klick, and D. L. Dexter, *Phys. Rev.* **103**, 1715 (1956); and J. J. Hopfield, *Phys. Chem. Solids* **10**, 110 (1959).

[378] Emission spectra particularly rich in detail are presented by W. J. Choyke and L. Patrick, *Phys. Rev.* **127**, 1868 (1962) (SiC); by Thomas and Hopfield[309]; and by A. T. Vink and C. Z. Van Doorn, *Phys. Letters* **1**, 332 (1962) (GaP).

[379] See D. S. McClure, ref. 14, Section 3b; R. M. Hochstrasser, *Revs. Modern Phys.* **34**, 531 (1962).

Indirect exciton decay can take place by a single-phonon process which is just the inverse of the indirect absorption process (Fig. 29). The lifetime against such a radiative decay is of the order of 10^{-6} sec at $T = 0°$ K, according to Toyozawa,[244] which is considerably shorter than lifetimes measured by Haynes et al.[380] in Si (6×10^{-5} sec) at 18° K.

b. Emission from Trapped Excitons

The luminescence from excitons bound to various impurities in semiconductors has only recently come under exhaustive investigation and will not be covered in this article, although the present section is a natural place for it. The principal objectives of these experiments seem to be studies of the impurities and the crystals themselves, using excitons as a probe, rather than vice versa. We shall restrict this discussion to luminescence from *self-trapped* excitons.

If an exciton is strongly self-trapped and does not lose all of its excitation energy by a thermal spike or resonance transfer to an impurity, then it is likely to luminesce with a large Stokes shift. The first observations of luminescence from very pure crystals which may have indicated this type of decay were by van Sciver and Hofstadter[381] and Teegarden,[382] in NaI and KI, respectively. In KI, irradiation at 5.9 ev leads to an emission at 3.3 ev which is relatively independent of the history and impurity content of the crystal. This would seem to indicate self-trapping with a large Stokes shift; it turns out[383] that a transient enhancement of the same emission is induced by simultaneous irradiation of the crystal with ultraviolet and red light. Teegarden believes that this can be explained on a model in which the uv photon creates a trapped hole and an F center; then the red light releases the electron to recombine with the hole. It is quite possible that the latter experiment is complementary to the original trapped-exciton fluorescence experiment, since a trapped hole might bind an electron into an excited state quite similar to the one created as a result of the exciton self-trapping process.

[380] J. R. Haynes, M. Lax, and W. F. Flood, *Proc. Intern. Conf. Semicond. Phys., Prague, 1960* p. 423 (1961).
[381] W. J. van Sciver and R. Hofstadter, *Phys. Rev.* **97**, 118 (1955); W. J. van Sciver, *ibid.* **120**, 1193 (1960).
[382] K. J. Teegarden, *Phys. Rev.* **105**, 1222 (1957).
[383] K. J. Teegarden and R. F. Weeks, *Phys. Chem. Solids* **10**, 211 (1959).

V. Summary

Before commenting on the present and future state of exciton research, we hasten to point out that it has not been possible to cover all areas of exciton research in this review, and to mention two in particular which have been left out. They are the theory of exciton production by fast charged particles and gamma rays, and the kinetics of electron-hole-exciton equilibrium phenomena. The former is of importance to the field of energy loss of particles in solids,[384] and to the study of mechanisms of energy transfer to luminescent centers in scintillators.[385] The kinetics problem occurs in connection with recombination studies in semiconductors, thermal conductivity theories (Section 12e), and studies of exciton statistics (Section 7a).

The electronic structure of the exciton can be regarded as quite well understood in most solids. In semiconductors Wannier exciton theory is capable of providing a very accurate description of all but the lowest exciton states, where first-order corrections due to exchange and polarization effects are present but small and understood in principle. Frenkel theory is successful in organic systems where excited states are indeed "tightly bound." Much is yet to be learned about intermediate cases (e.g., the alkali halides and rare gas solids), where first-order corrections to the Wannier model become very large or, in terms of the Frenkel model, overlapping of excited state wave functions in different cells becomes large. The exciton is understood in principle, in these cases, however; it is only the details of the wave functions and fairly precise excitation energy calculations which are now lacking. We point out here that the excited electronic structure of mixed crystals has received little theoretical attention, and that several experiments on absorption edge shifts as a function of composition remain to be explained.[386]

---

[384] See, in particular, F. Seitz.[133] A typical recent calculation bearing on this problem is by J. Lory, *Compt. Rend.* **250**, 3622 (1960).

[385] See, e.g., R. B. Murray and A. Meyer, *Phys. Rev.* **122**, 815 (1961).

[386] See, e.g., W. Martienssen, *Proc. Intern. Conf. Semicond. Phys., Prague, 1960* p. 760 (1961); E. Taglauer and W. Waidelich, *Z. Physik* **169**, 90 (1962).

Exciton dynamics, on the other hand, is characterized by a certain amount of disorganization in the existing literature. In the matter of exciton-phonon coupling constants, for example, one finds only isolated calculations for particular crystals and particular types of excitons. We have attempted to at least juxtapose these calculations here, but it has not been possible to go deep into the details of each case in order to establish ranges of validity of all their approximations. Such a study would be welcome.

In other areas of exciton dynamics, we believe that the next few years will see greater application of the polariton concept, especially as it bears on the details of absorption edge shapes. The primary outstanding problem in edge shapes is Urbach's rule, as discussed in Section 10c. We suggest that anomalous waves should be sought near phonon absorption bands as well as in exciton bands. The argument for the existence of "anomalous reststrahl waves" is exactly the same as that given for anomalous light waves near exciton bands. The possibility of observing phonon anomalous waves might be enhanced because of the existence of long phonon mean free paths at low temperatures.

Although exciton energy transfer occurs in certain molecular crystals by a hopping process, energy transfer by means of truly delocalized wave packets has not in our opinion been conclusively demonstrated in any solid as yet. Photoconductivity can be induced by absorption in the lowest exciton band of many solids, and can be understood on the basis of exciton dissociation or exciton ionization of defects. When the exciton binding energy is large (e.g., alkali halides) this photoconductivity in pure crystals is accordingly quite small. An exception to this rule is the case of organic crystals, where one would expect very little photoconduction as a result of exciton-band excitation. It will be most helpful if photoconductivity studies are carried out on the solid rare gases, solids which combine many characteristics of alkali halide and molecular crystals. On the basis of the great similarity between the rare gas and alkali halide absorption spectra,[64] one would guess that photoconductivity would be small in the lowest exciton absorption peaks.

An exceedingly valuable tool for the study of exciton dynamics and statistics will be the tunable laser, especially one operating in the ultraviolet. When such a device is available, its high photon flux can be used to heavily populate exciton states and enable one to search for nonlinear effects such as collision ionization, or to reach a density at which Bose-Einstein condensation might take place. The exciton itself, as a dynamical system, will remain an intriguing object of study for the next several years.

Author Index

Numbers in parentheses are reference numbers and indicate that an author's work is referred to although his name is not cited in the text.

A

Abramson, E., 170, 178
Adams, E. N., 81
Agekyan, V. T., 86
Agranovich, V. M., 119, 134(206), 135(206), 140, 144(247), 152, 157, 165, 174(247, 265)
Allen, J. W., 79
Anderson, P. W., 100
Ansel'm, A. I., 137, 143, 144(242, 243), 148(242, 243)
Aoyagi, K., 180
Apfel, J. H., 176
Apker, L., 176, 179, 186
Arakian, P., 170, 178

B

Baldini, G., 33, 34(64), 70, 122(64), 130(64), 192(64)
Balkanski, M., 47, 90, 99
Bardasis, A., 5, 100
Bardeen, J., 100, 123, 142, 144, 148, 161(213), 174(254)
Barriol, J., 52
Bassani, F., 73, 183
Baumeister, P. W., 3, 52, 122(80)
Biellman, J., 57, 130
Biem, W., 149
Bierman, A., 172
Blakney, R. M., 175
Blatt, F., 123, 161(213)
Blatt, J. M., 91
Boer, K. N., 91
Bogoliubov, N. N., 100
Bonch-Bruevich, V. L., 27
Born, M., 25, 27, 28, 64, 107, 109(193), 110(50), 137
Bouckaert, L. P., 27, 29
Bradburn, M., 25, 28
Brandt, W., 91
Braunstein, R., 136
Brillouin, L., 105, 106(192)
Brodin, M. S., 130, 167, 168
Broude, V. L., 32
Brout, R., 147
Brown, F. C., 58, 161, 163(100)
Brueckner, K. A., 95, 97(174)
Bulyanitsa, D. S., 86
Burckel, J., 57
Busch, G., 184
Button, K., 82

C

Cardona, M., 85
Cario, G., 171, 182(306)
Caris, J. C., 170, 178
Casella, R. C., 91
Cauchois, Y., 101
Chadderton, L. T., 87
Cherepanov, V. I., 115, 124
Choi, S., 170, 178
Choyke, W., 58, 163(99), 189
Clark, C. D., 163
Coelho, R., 130

Cohen, M. H., 25, 29
Condon, E. U., 62, 134, 135(236)
Cooper, L. N., 100
Craig, D. P., 32

D

Davydov, A. S., 4, 29, 141, 149, 165
Davydov, V. I., 144
Deb, S. K., 58
de Boer, J., 64
Delbecq, C. J., 183, 187
Demidenko, A. A., 25
des Cloizeaux, J., 47
Devyatkova, E. D., 184
Dexter, D. L., 4, 21, 22(40, 41), 33(41), 67, 68(123), 70(123), 73(123), 90(40), 112, 118, 128(10, 123), 129(10), 131, 134, 135(235), 155, 156, 171, 175, 179, 182(306), 186, 189
Dianov-Klokov, V. I., 135
Dieke, G. H., 134
Dietz, R. E., 58
Dimmock, J. O., 44, 58, 82
Dresselhaus, G., 5, 42, 46, 119
Dressler, K., 5, 32, 33(22)
Druzhinin, V. V., 124
Dutton, D. B., 55, 56, 63, 66(109), 69(109), 72, 111, 130(109), 155(86), 180(109)
Dykman, I. M., 64, 66, 129(119), 181, 182

E

Eby, J. E., 63, 66(109), 69(109), 72, 111, 130(109), 180(109), 187
Edwards, D. F., 82, 83, 128(149)
Ehrenreich, H., 130
El Komoss, S. G., 129, 130
Elliott, R. J., 5, 45, 53, 79, 81, 87(70), 119, 123(21), 124(70), 126, 127, 147, 152, 158(21)
El-Sayed, M. A., 188
Englert, F., 50, 96

F

Fan, H. Y., 58
Fano, U., 27, 108, 164

Ferguson, J. N., 180
Firsov, Yu. A., 137, 143, 144(242, 243), 148(242, 243)
Fischer, F., 66, 70, 72, 73
Flood, W. F., 190
Foerster, T., 171, 182(306)
Fox, D., 25(44)
Franck, J., 62, 169, 171, 182(306)
Franz, W., 157
Frenkel, J., 1, 21, 22, 144, 169, 171(4), 181
Friedel, J., 98
Frohlich, H., 93, 142
Fuchs, R., 183

G

Galischev, V. S., 115, 124
Genkin, G. M., 148
Gilleo, M. A., 155, 158(278)
Ginzburg, V. L., 165, 168
Givens, M. P., 63, 130(110), 135
Gold, A., 34, 148, 174(262)
Goldstone, J., 95, 97(174)
Goodman, B., 66, 90, 170(121), 177
Gorban', I. S., 162, 167
Goto, F., 70
Gribnikov, E. S., 172
Griffith, H. O., 33
Gross, E. F., 5, 28, 52, 53, 55, 56(53), 58, 74(53, 132), 77, 79, 82, 83, 86, 121(209), 124, 126(53), 130(16), 169, 176, 189
Grun, J. B., 122, 130, 162
Gurney, R. W., 64
Guseva, G. I., 99

H

Haga, E., 141
Haken, H., 6, 50, 53, 93, 94, 154, 172, 173
Hall, L. H., 123, 161(213)
Harbeke, G., 85
Hartman, P. L., 33, 63, 66(109), 69(109), 73(109), 130(109), 180
Haupt, U., 155
Hayashi, M., 52
Haynes, J. R., 172, 182(309), 190

AUTHOR INDEX

Hebb, M. H., 179
Heller, W. R., 21, 22, 27, 35(39), 90(40), 92, 141(39), 179, 186
Hensley, E. B., 130
Herring, C., 183
Herzfeld, K. F., 4
Hilsch, R., 4, 62, 66(6, 118), 70, 72, 73
Hirai, M., 182, 186
Hobbins, P. C., 32
Hochstrasser, R. M., 189
Hoffman, R., 119,
Hofstadter, R., 190
Hopfield, J. J., 27, 28(49, 53), 47, 55, 56(53), 58, 74(53), 77, 78, 79(53, 89), 82, 83, 84, 86, 121(209), 124(53), 126, 131, 152, 157, 163(88), 168, 172, 180, 182, 189
Horie, C., 98, 99
Hrivnak, L., 51, 175
Huang, K., 27, 107, 109(193), 110, 137
Hubbard, J., 108

I

Ikezawa, M., 186
Imai, I., 57, 155(91)
Inchauspé, N., 70, 72(128), 130(128), 183
Izuyama, T., 90, 100

J

Jahoda, F. C., 112
Jakobsen, M. A., 55
Jarnagin, R. C., 178
Johansson, P., 101
Johnson, E. J., 58
Jossem, E. L., 101

K

Kachlivskii, Z. S., 174
Kaifu, Y., 186
Kallman, H., 176
Kanskaya, L. M., 86
Kaplianskii, A. A., 53, 86, 124, 176
Kargapolov, Yu. A., 115, 124
Karryeff, N. A., 52
Katsuki, K., 52
Katul, J., 176

Keffer, F., 25, 29
Keldysch, L. V., 157
Kepler, R. G., 178
Keyes, R. W., 81
Kittel, C., 90
Kiyono, S., 101
Kleiner, W. H., 82, 87
Klemm, W., 64
Klick, C. C., 171, 182(306), 189
Knox, R. S., 5, 33, 34, 70, 72(128), 130, 148, 163(222), 174(262), 187
Kobayashi, K., 155
Kohn, W., 13, 44, 59, 95
Konobeev, Yu. V., 140, 144(247), 152, 157, 174(247, 265)
Konstantinov, O. V., 124
Korenblit, L. L., 79, 81
Korolyuk, S. L., 172, 184(312)
Koster, G. F., 13, 27, 49(52), 61
Krochuk, A. S., 130
Kroger, F. A., 188, 189
Kuang-Yin, C., 53, 86(82)
Kuhn, H., 62
Kuwabara, G., 180

L

Lamb, W., 80
Lambe, J., 189
Lampert, M. A., 172, 182(309)
Lax, B., 82, 85, 128, 149
Lax, M., 61, 146, 177, 190
Lazazzera, V. J., 82, 83, 128(149)
Lee, T. D., 93
Leurgans, P. J., 144, 148, 174(254)
Lipnik, A. A., 177, 179
Löwdin, P. O., 11, 33, 34(31)
Lory, J., 191
Loudon, R., 79, 81, 126, 127, 136, 162
Low, F. E., 93, 173
Lubchenko, A. F., 165, 168
Luttinger, J. M., 44, 59 (68)
Lynden-Bell, R., 33

M

McClure, D. S., 5, 30(14), 32, 119, 188, 189
McConnell, H. M., 33
Macfarlane, G. G., 58, 83, 158(97, 98)

McLean, T. P., 5, 58, 83(97), 111, 158(19, 97, 98), 162, 163(19)
Mahr, H., 158
Marcus, A., 21, 22(39), 27, 35(39), 92, 141(39)
Martienssen, W., 148, 155, 191
Masumi, T., 58, 161, 163(100)
Medvedev, V. S., 32
Meijer, H. J. G., 188, 189
Merrifield, R. E., 90, 175
Meyer, A., 191
Milgram, A. A., 63, 130(110), 135
Miyakawa, T., 135
Moser, F., 155
Moskalenko, S. A., 45, 51, 59, 91
Mott, N. F., 4, 5, 49, 64, 69, 98, 101
Murray, R. B., 191
Muto, T., 69, 101(126)

N

Nakai, Y., 180
Nelson, J. R., 33, 63, 66(109), 69(109), 73(109), 130(109), 180(109)
Nieman, G. C., 188
Nikiforov, A. E., 124
Nikitine, S., 5, 52, 121, 122, 130, 154, 162, 172
Northrup, D. C., 170
Noskov, M. M., 152
Novikov, B. V., 176
Nozieres, P., 90, 96, 98, 99

O

Obreimov, I. W., 4
O'Bryan, H. M., 63, 66(109), 69(109), 73(109), 130(109), 180(109)
Oen, O. S., 66, 90, 170(121), 177
Ohta, T., 154
Okamoto, Y., 130
Okuno, H., 69, 101(126)
Olness, D., 178
Onaka, R., 63, 130(110)
Osaka, Y., 70
Osaka, Y. S., 70
Ovander, L. N., 135
Overhauser, A. W., 4, 66, 128(11)
Oyama, S., 69, 101(126)

P

Pappert, R. A., 66, 128(122), 129(122)
Paranichev, V. N., 172
Parratt, L. G., 101
Parsons, R. B., 87
Pastrnyak, I., 162
Patrick, L., 58, 163(99), 189
Pavinskii, P. P., 54, 82
Peierls, R. E., 1, 22, 33(3), 137(3), 144, 146, 171(3)
Pekar, S. I., 27, 163, 165, 167, 168, 181, 182
Pelzer, H., 93, 142(252)
Perny, G., 57, 154
Petersen, R., 130
Peterson, D. L., 141, 144(250)
Philipp, H. R., 112, 176, 177(334), 179, 180(337), 186(337)
Pikus, G. E., 172, 184(311)
Pines, D., 93, 96, 97, 98
Piskovoi, V. N., 25
Pohl, R. W., 4, 62, 66(6)
Pooley, D., 33
Popov, Yu. M., 188
Portis, A. M., 176
Power, M., 55, 163(88)
Pratt, G. W., Jr., 95
Prikhot'ko, A. F., 4, 32
Pringsheim, P., 183

Q

Quarrington, J. E., 58, 83(97), 158(97, 98)

R

Radchenko, R. V., 115, 124(202)
Radkowsky, A., 154
Raimes, S., 97, 98(177)
Rashba, E. I., 45, 149, 172
Razbirin, B. S., 28, 55, 56(53), 74(53), 79(53), 82, 83, 121(209), 124(53), 126(53)
Redfield, D., 157
Reiling, G. H., 130
Reinov, N. M., 74, 77(132)
Reiss, R., 57, 121, 130
Reitz, J. R., 11, 33(32)

Rice, S. A., 170, 178
Rimmer, M. P., 112
Ringeisen, J., 130
Roberts, V., 58, 83(97), 158(97, 98)
Robinson, G. W., 188
Rollefson, G., 62
Roth, L. M., 82, 87, 95, 128, 149
Rukhadze, A. A., 165, 168
Ryvkin, S. M., 176

S

Saum, G. A., 130
Schiff, L. I., 78, 87, 113, 114(201), 126, 131, 173
Schneider, E. G., 63, 66(109), 69(109), 73(109), 130(109), 180(109)
Schneider, M., 184
Schnepp, O., 5, 32, 33(22)
Schottky, W., 50, 93, 142
Schrieffer, J. R., 5, 100
Schulman, J. H., 171, 182(306)
Schultz, T. D., 143
Seitz, F., 14, 21, 27, 33(38), 35(36), 64, 77, 169, 171, 175, 176, 182, 191
Selivanenko, A. S., 175, 188
Shirkov, D. V., 100
Shmushkevich, B., 144
Shockley, W., 7, 21
Shortley, G. H., 62, 134, 135(236)
Shur, M. Ya., 152
Sibilev, A. I., 82
Siegfried, J. G., 63, 66(109), 69(109), 73(109), 130(109), 180(109)
Sieskind, M., 52, 121, 122, 130, 154, 162
Silin, V. P., 168
Silver, M., 176, 178
Simpson, O., 170, 182, 188(353)
Simpson, W.T., 141, 144(250)
Slater, J. C., 7, 12, 21, 61
Smoluchowski, R., 27, 29
Snyder, H., 126
Sobolev, V. V., 58
Sternlicht, H., 33
Strizhevskii, V. L., 165
Stumpf, H., 182
Sturge, M. D., 58
Svirskii, M. S., 91
Swicord, M., 178

T

Taft, E. A., 112, 176, 179, 180(337), 186
Taglauer, E., 191
Takeuti, Y., 61
Taluts, G. G., 99
Taylor, J. W., 180
Teegarden, K. J., 63, 66(109), 69(109), 72, 111, 130(109), 171, 180, 186, 187, 190
Teller, E., 169
Thomas, D. G., 27, 28 (49, 53), 56(53), 58, 74(53), 77, 78, 79(53, 89), 82(53, 89), 83, 84, 86, 87, 121(209), 124(53), 126, 129(93), 163(88), 172, 182, 189
Thouless, D. J., 8, 87
Tibbs, S. R., 68, 70(124)
Timofeev, V. B., 162, 167
Tippins, H. H., 58, 161, 163(100)
Tolmachev, V. V., 100
Tolpygo, K. B., 45, 51, 59, 66, 110
Tomasevich, O. F., 66
Tomiki, T., 155
Tomura, M., 186
Toyozawa, Y., 6, 90, 92, 135, 137, 145, 148(27), 149, 150, 151(245), 152, 156, 170, 171(27), 176, 179, 188(244), 189, 190
Trlifaj, M., 171, 179, 182
Tsekvava, B. E., 165
Tsertsvadze, A. A., 66, 129(119), 183
Tulub, A. V., 173, 175(317)
Tutihasi, S., 130, 153, 155(230)

U

Ueta, M., 5, 52(17), 130(17), 182, 186
Urbach, F., 154, 155, 158

V

Van Doorn, C. Z., 189
van Sciver, W. J., 190
Van Vleck, 146
Varfalomeev, A. V., 124
Varsanyi, F., 134
Vink, A. T., 189
von Hippel, A., 4, 64, 154(8)
Vonsovskii, S. V., 91

W

Waidelich, W., 191
Wannier, G. H., 1, 7(5), 176
Wardzynski, W., 87
Watanabe, H., 186
Wauk, M. T., 188
Weeks, R. F., 190
Weller, W., 149
Wheeler, R. G., 44, 58, 82
Wigner, E. P., 27, 29
Winston, H., 33, 119
Wolf, H. C., 5, 30(15), 188
Wolff, K. L., 4

Y

Yaet, Y., 81
Yatsiv, S., 25(44)
Yoffe, A. D., 87
Yokota, I., 141
Yuster, P. H., 183, 187

Z

Zahlan, A. B., 176
Zakharchenya, B. P., 74, 77(132), 79, 82, 86, 124
Zaslavskaya, I. G., 154
Zhilich, A. G., 54, 115, 124
Zhuze, U. P., 176
Zienau, S., 93, 142(252)
Zinngrebe, H., 130, 153(230), 155(230)
Zollweg, R. J., 130, 171(334), 176
Zverev, L. P., 152
Zwerdling, S., 82, 128, 149

Subject Index

A

Absorption coefficient,
  definition of, 105
  energy dependence of, 144–163
  in exciton continuum, 123, 161–162
  integrated, definition of, 116
  on Lorentz model, 107
  measurement of, 111
  semiclassical expression for, 116
Absorption edge,
  shape in presence of exciton and magnetic field effects, 127–128
Absorption of light,
  physical basis of, 122, 133–134, 136–137, 145–146
Alkali azides,
  exciton levels in, 58
Alkali halides, see also individual solids
  band structure of, 71–72, 74
  doublet intensities in, 130
  exciton levels in, 63
  exciton mean free paths in, 148, 174, 175
  photoconductivity in, 180
  photoemission from, 176
  transport of energy by excitons in, 186–187
"Allowed" transitions,
  direct, oscillator strengths for, 121
Alpha band, 183
Anomalous waves, see Spatial dispersion
Anthracene,
  Davydov splitting in, 32
  exciton diffusion length in, 182, 188
  exciton mean free path in, 174
  photoconductivity in, 178
  spatial dispersion in, 167
Argon, solid,
  absorption line in, 34
  exciton band structure of, 35
  Frenkel excitons in, 33–37
Asymmetry,
  in absorption lines, causes of, 152

B

Barium oxide,
  photoemission from, 176, 177
Benzene,
  Davydov splitting in, 32
Beta band, 183
Bloch representation,
  of excited states of a crystal, 14–15
  of one-electron states, 10
Bose–Einstein condensation of excitons, 90–91, 134
Bound exciton, see Trapped exciton

C

Cadmium selenide,
  exciton levels in, 58
  strain effects in, 87
  Zeeman effect in, 82
Cadmium sulfide,
  band structure of, 55
  edge emission from, 189
  energy transport in, 187
  exciton levels in, 55–57
  field ionization of excitons in, 79
  indirect transitions in, 163
  longitudinal excitons in, 28, 57
  magnetic effects on intensities in, 124–126
  magnetostark effect in, 85–86
  photoconductivity in, 78
  Stark effect in, 78

Urbach's rule in, 155
Zeeman effect in, 82
Cadmium telluride,
 strain effects in, 87
Carbon monoxide, solid,
 Davydov splitting in, 32
Central-cell corrections, see Wannier exciton, first-order corrections in
Copper halides,
 exciton levels in, 130
Crystal Hamiltonian, see Hamiltonian
Cuprous oxide,
 absorption spectrum of, 3
 asymmetry of $n = 2$ line in, 152
 band structure of, 54
 exciton levels in, 52–55, 130
 field ionization of excitons in, 77, 79
 first-order exciton energies in, 45
 magnetic effects in, 82
 photoconductivity in, 176
 quadrupolar transitions in, 124
 spatial dispersion in, 167, 168
 Stark effect in, 77
 strain effects in, 86–87
 temperature dependence of band gap in, 154

D

Davydov splitting, 29–33
 definition, 32
 observation of, 32
Deformation potential theory,
 matrix elements in, 142
Detectors, of excitons, 185
Diamond,
 indirect transitions in, 163
Dielectric constant,
 choice of, in Wannier theory, 49–52
 complex, 104
 on Lorentz model, 107
Diffusion of excitons, 171
Dipole lattice sums, 25, 29
Direct transitions,
 theory of, 112–134
Dislocations,
 effect of, on absorption edge, 175
 scattering of excitons by, 175

Dissociation of excitons, see also Photoconductivity
 by collisions between excitons, 178
 mechanisms of, 170–171, 177–181
 by phonons, 177
 by static imperfections, 178–180
Doublet splitting, see Spin-orbit interaction

E

Edge emission, 189
Effective mass,
 of Frenkel exciton, 28, 35
 of Wannier exciton, 38, 85, 148
Electric field effects, see also individual solids
 on diffusion of excitons, 172
 ionizing, 79
 Stark effect in hydrogen, 78
 theory of, 74–79
Electron-transfer model of the exciton, 61–67, 70–74
 connection with the Wannier model, 65
Electronic polaron, 92
Emission, see Radiative decay
Energy transport, see Transport of energy by excitons
Excitation model of the exciton, 67–74
 connection with the Wannier model, 67–68
Exciton–phonon interaction, see also specific exciton models
 general theory of, 137–140
Exciton-photon interaction, see also specific exciton models
 in second quantization, 131–132
 semiclassical theory of, 113–117
Exciton representation, 17–20
Exciton rydberg,
 definition of, 39
 experimental determination of, 51, 52–58
 size of, relative to band gap, 100
Extinction coefficient,
 definition of, 104

F

F centers,
 ionization of, by excitons, 179

SUBJECT INDEX

Factor group splitting, *see* Davydov splitting
Field ionization, *see* Electric field effects
"Forbidden" transitions,
  direct, oscillator strengths for, 121
Frenkel exciton, 21–37
  eigenvalues and eigenfunctions, 21–24
  phonon interaction with, 140–141
  photon interaction with, 118–119

G

g-value, exciton, 83
Gallium antimonide,
  exciton levels in, 58
Gallium arsenide,
  energy transport in, 187
  exciton levels in, 58
Germanium,
  band gaps in, 83–85
  direct transitions in, 178
  exciton levels in, 58, 162
  exciton lifetime in, 149
  exciton mean free paths in, 148
  indirect transitions in, 162
  magnetic effects in, 82–85
  strain effects in, 87

H

Hamiltonian,
  crystal, 7–8, 137, 139
    in presence of electric field, 74, 77
    in presence of electromagnetic field, 113
    in presence of magnetic field, 80–83
  effective, for electron-hole pair, 37, 51, 91–97
  of excess charge in crystal, 92
  of idealized model of an insulator, 9
  phonon, 138
"Hopping," *see* Transport of energy by excitons
Hydrogenic exciton, *see* Wannier exciton

I

Index of refraction,
  of "anomalous waves," 164, 166

  complex, 104
  real part of, equations determining, 163
Indirect transitions, 147, 158–163
  in emission, 190
  transition probabilities for, 159, 161–162
Indium antimonide,
  thermal conductivity of, 184
Intermediate-radius excitons,
  eigenfunctions and eigenvalues, 60–61
  photon interaction with, 128–130

K

Krypton, solid,
  doublet intensities in, 130

L

Lead iodide,
  exciton levels in, 57
  Urbach's rule in, 155
Lead telluride,
  thermal conductivity in, 184
Lifetime of exciton states,
  against phonon dissociation, 177
  against phonon scattering, 147
Lithium fluoride,
  exciton levels in, 130, 135
Local field effects, 117, 118–119
Localized representation, *see also* Wannier functions
  for excited states of a crystal, 16
  for one-electron states, 11–13
Longitudinal excitons, *see also* Transverse excitons and Symmetry properties of excitons
  definition of, 26
  excitation of, theory, 114–115
  in the Frenkel case, theory of, 24–29
  in the Lorentz model, 108
  observation of, 28, 57
  in the Wannier case, theory of, 45–46
Lorentz model, 106
  spatial dispersion in, 164–167
Lorentzian shape,
  definition of, 107
  resulting from weak-coupling theory, 151

Luminescence, exciton, *see* Radiative decay

M

Magnetic field effects, *see also* individual solids
  in crystals lacking inversion symmetry, 124–126
  on diffusion of excitons, 172
  on eigenstates and eigenvalues, 79–85, 126–127
  on transition probabilities, 124–128
  in two-particle effective Hamiltonian, 80, 83
Magnetostark effect, 85–86
Mass, exciton, *see* Effective mass or Reduced mass
Maxwell equations, 103
Mean free path, exciton,
  definition of, 174
  due to phonon scattering, 174
Mercurous iodide,
  exciton levels in, 57, 130
  photoconductivity in, 176
Molecular crystals,
  exciton states in, 29–33
Motion of excitons, *see also* Transport of energy by excitons
  definitions of, 169–170
  by diffusion, 182
Multiplicity, *see also* Spin-orbit interaction
  on electron-transfer model, 66, 70, 73
  on excitation model, 70

N

Naphthalene,
  Davydov splitting in, 32

O

One-electron approximation, 7–20
  in excited states of solids, 13–20
Optical density,
  definition of, 111

Organic crystals, *see also* individual crystals
  emission from, 189
  energy transport in, 187–188
  photoconductivity in, 176
Oscillator strength, *see also* Absorption coefficient
  approximations in calculations of, 118
  definition of, 117
  for direct transitions, 117–122
  relation to absorption coefficient, 117
Overlap of wave functions,
  corrections for, 33–36
  in second-quantized formalism, 90
Oxygen, solid,
  two-exciton processes in, 135

P

Phonons,
  acoustic, scattering of Bloch electrons by, 142
    scattering of excitons by, 143
  boson operator formalism for, 139
  effects of, 136–163
  normal coordinate expression for, 138
  optical, scattering of Bloch electrons by, 142
    scattering of excitons by, 144
  role of, in indirect transitions, 158–163
Photoconductivity, *see also* Dissociation of excitons; and individual solids
  discussion of, 180
  zero contribution of exciton to, 176
Photoemission, 176–181
  from F centers, 179, 186
Plasmon,
  exciton's relation to, 99–100
  qualitative discussion of, 97–98
Polariton, 133
  as a carrier of energy, 133, 184, 185
  scattering cross section for, 173–174
Polarization of light,
  dependence of optical absorption on, 114–117, 119
Potassium bromide,
  absorption spectrum of, 2
  doublet intensities in, 130

SUBJECT INDEX

photoconductivity in, 180
Urbach's rule in, 155
Potassium chloride,
photoconductivity in, 180
Urbach's rule in, 155
Potassium iodide,
emission from trapped excitons in, 190
photoconductivity in, 180
Urbach's rule in, 155

R

R center,
bleaching of, 182
Radiative decay,
of excitons, 170, 188–190
of "free" excitons, 188–189
of trapped excitons, 190
Radius,
exciton, 3, 4, 49, 51, 59
Rare gases, solid,
exciton mean free paths in, 148, 174
Recoil of exciton,
during photon emission, 169, 189
Recombination, electron-hole,
importance of, in organic crystals, 178
with phonon assistance, 177
Reduced mass, 38
Refractive index, see Index of refraction
Renormalization of charges in a dielectric medium, 97
"Retardation" effects,
in classical theory of dispersion, 107–111
in quantum theory of dispersion, 130–134
Rubidium bromide,
doublet intensities in, 130
Rydberg, see Exciton rydberg

S

Scattering of excitons,
cross section for, 173
by dislocations, 175
by phonons, 174
by point defects, 174–175
by surfaces, 175–176

Second quantization, notation,
for excitons, 87–90
for phonons, 139
for photons, 131–132
Selenium, amorphous,
Urbach's rule in, 155
Self-trapping, of excitons, 144, 171, 181–182, see also Trapping
criteria for, 181, 182
Semiconductors, see also individual solids
energy transport by excitons in, 187
Shifts with temperature of absorption lines and edges, 147, 152–154
Silicon,
emission from, 190
exciton levels in, 58
indirect transitions in, 158, 163
Silicon carbide,
exciton levels in, 58
Silver bromide,
doublet intensities in, 130
indirect transitions in, 163
Urbach's rule in, 155
Silver chloride,
doublet intensities in, 130
exciton levels in, 58
indirect transitions in, 161, 163
Urbach's rule in, 155
Silver iodide,
exciton levels in, 57
Sodium chloride,
oscillator strengths in, 129
Sodium iodide,
emission from trapped excitons in, 190
Spatial dispersion,
in the case of excitons, 163–168
of phonons, 192
Spin-orbit interaction,
in beta band, 183
in crystal Hamiltonian, 8
in cuprous oxide, 54–55
effect on alkali halide spectra, 62–63, 71–73, 130
on exciton g values, 83
on longitudinal-transverse splitting, 29
matrix elements of, 20

"Spin-spin" interaction, *see* Wannier exciton, first-order corrections in
Stark effect, *see* Electric field effects
Statistics of the exciton, 89–91
Strain field effects,
  on exciton levels, 86–87
  on one-electron bands, 86
Strong coupling,
  definition, in exciton-phonon problems, 151
Superconductors,
  excitons in, 100
Surface excitons, 175–176
Surfaces,
  effects of, 111, 176
Symmetry properties of excitons, *see also* Longitudinal excitons and Transverse excitons
  in cubic crystals, 27
  regarded as an angular momentum coupling problem, 47
  translational, experimental verification of, 86

T

Tetracene,
  surface excitons in, 176
Thallium halides,
  exciton levels in, 130
Thallous chloride,
  absorption spectrum of, 153
  Urbach's rule in, 153, 155
Thermal conductivity,
  due to excitons, 172, 183–184
Thermal spike,
  in color center kinetics, 182
  definition of, 171
Thermalization of excitons, 170, 189
Tightly-bound exciton, *see* Frenkel exciton
Transport of energy by excitons,
  general question of, 169, 184–186
  by hopping, 184–185
  as thermal current, 172, 183–184
  by wave packets, 184–186
Transverse excitons, *see also* Longitudinal excitons and Symmetry properties of excitons
  coupling with photons, 115
  definition, 26
  in Frenkel case, theory of, 24–29
  on the Lorentz model, 108–109
  in Wannier case, theory of, 45–46
Trapped exciton,
  in alkali halides, 183
  identity with excited states of effects, 182–183
Trapping, *see also* Self-trapping
  of excitons, 172, 181–183
Two-electron transitions, 134–136
Two-photon absorption, 136

U

U centers,
  as exciton detectors, 186
Urbach's rule, 152–158
  statement of, 154
  table of cases of, 155
  theories of, 155–158

W

Wannier exciton, 37–59
  effective electron-hole interaction in, 50
  eigenvalues and eigenfunctions of, 37–49
  first-order corrections in, 44–46
  lifetime of, against phonon scattering, 148
  normalization of envelope function of, 42
  phonon interaction with, 141
  photon interaction with, 119–121
Wannier functions,
  definition of, 12
Wave vector of light,
  complex, 104
  order of magnitude of, 117
Weak coupling,
  definition, in exciton-phonon problems, 150–151

Weakly-bound exciton, *see* Wannier exciton

X

X-ray excitons, 100–101
Xenon, solid,
 exciton levels in, 70

Z

Zeeman effect, *see* Magnetic field effects
Zinc oxide,
 exciton levels in, 58
 longitudinal excitons in, 28
Zinc selenide,
 exciton levels in, 58

QC
176
K58
1963

MAR 19 1964